World Survey of Climatology Volume 8

CLIMATES OF NORTHERN AND EASTERN ASIA

World Survey of Climatology

World Survey of Climatology Volume 8

Climates of Northern and Eastern Asia

edited by H. ARAKAWA

Meteorological Research Institute
Tokyo (Japan)

ELSEVIER PUBLISHING COMPANY Amsterdam-London-New York 1969

ELSEVIER PUBLISHING COMPANY
335 Jan van Galenstraat
P.O. Box 211, Amsterdam, The Netherlands

ELSEVIER PUBLISHING COMPANY LTD.
Barking, Essex, England

AMERICAN ELSEVIER PUBLISHING COMPANY, INC.
52 Vanderbilt Avenue
New York, New York 10017

Library of Congress Card Number: 68–12480

Standard Book Number 444–40704–9

With 137 illustrations and 36 tables

Printed in The Netherlands

World Survey of Climatology

List of Contributors to this Volume:

H. ARAKAWA
Meteorological Research Institute,
Mabashi, Suginami-ku,
Tokyo (Japan)

V. F. BALAGOT
Weather Bureau Forecasting Center,
Manila International Airport,
Manila (Philippines)

J. F. FLORES
Weather Bureau Forecasting Center,
Manila International Airport,
Manila (Philippines)

M. SUKANTO
Meteorological and Geophysical Service,
Djakarta (Indonesia)

S. TAGA
Climatic Section,
Japanese Meteorological Agency,
Tokyo (Japan)

I. E. M. WATTS
Royal Observatory,
Kowloon (Hongkong)

Preface

This book, the eighth volume of the World Survey of Climatology series, is one of the two volumes of this series dealing with the climates of Asia, exclusive of the U.S.S.R. (Because of the enormous amount of territory encompassed by the Soviet Union, a separate volume for this series is being prepared which will handle both the European and Asian areas of the U.S.S.R.) This present volume deals with China, the largest country of continental Asia, and five other Asian countries which have shorelines on the Pacific Ocean or its marginal seas: Japan, Korea, Taiwan, the Philippines and Indonesia. Inclusion of Indonesia in this volume will cause a partial overlapping with the second volume of this Asian series, *Climates of Southern and Western Asia*, which will include climatological reports on Vietnam, Laos, Thailand, Cambodia, Malaya, Burma, India, Pakistan, the Himalayan countries and the Near East.

The contributors to this volume can be considered authorities in their particular areas. I. E. M. Watts was Director of the Royal Observatory, Hong Kong, while preparing his manuscript on China and Korea. S. Taga is chief of the Climatic Section of the Japanese Meterological Agency at Tokyo. J. F. Flores and V. F. Balagot work for the Weather Bureau at Manila and M. Sukanto is Director of the Meteorological and Geophysical Service at Djakarta.

Therefore, the data published in this volume can be considered to be authoritative and up-to-date information of these areas of the world. Hopefully it will be of service to experts, as well as students, of climatology and related fields.

H. ARAKAWA

Contents

Chapter 2. CLIMATE OF JAPAN
by H. ARAKAWA and S. TAGA

Chapter 3. CLIMATE OF THE PHILIPPINES
by J. F. FLORES and V. F. BALAGOT

Contents

Chapter 4. CLIMATE OF INDONESIA
by M. SUKANTO

Contents

Chapter 1

Climates of China and Korea

I. E. M. WATTS

Introduction

The aim of this contribution is to provide, for as large a public as possible, a systematic account of the climates of central East Asia. For those who require further detail it has been necessary to include references to source material, but it is believed that these are not sufficiently numerous to make reading difficult.

As far as possible, average monthly or seasonal conditions have been explained with reference to the major atmospheric pressure systems and to the daily weather patterns which most commonly occur in each season. This was considered necessary because, with a country extending over 35° of latitude, both arctic and tropical influences play some part in producing large variations in climate.

The climates of Korea, Taiwan, Sinkiang and Szechuan have been dealt with separately. Elsewhere the information provided in the different sections on wind, precipitation and temperature should be sufficient to furnish a general description of average climatological conditions in any particular province or even smaller area. More comprehensive climatological statistics for a number of observing stations are given in Tables XXVII–LXVII. The section on regional climatological classifications posed a special problem. It seemed that, if this aspect was to be attempted at all, it had to be presented in full detail. Though this section is perhaps not completely of general interest, lengthy captions have been appended to the diagrams so that the various classification systems can be understood without the necessity of referring to the original papers or to other textbooks.

It is hoped that the use of metric units will not create too much of a problem for those who are only accustomed to the English system of units. The decision to use metric units was not made lightly. Throughout East Asia, rainfall is measured in millimetres and visibility in metres or kilometres, and the Congress of the World Meteorological Organization recently agreed that metric units should be adopted universally for meteorological observations. Some simple conversions from one system to the other are given in Table LXIX.

Place-names

In studying the climatology of Central and East Asia many problems arise in connection with place-names. Uncomplicated standard Chinese ideographs can be read by all Chinese people regardless of birthplace or dialect and are also reasonably intelligible to many of the people in neighbouring countries. However, in converting a place-name into

English there are two avenues. In a limited number of cases it is customary to translate the actual meaning of the name into English regardless of the sound. For example *Hwang Ho* 黄河 can be translated as "Yellow River" and the Korean or Chinese *Do* as "Island". *Shan* 山 can usually be translated as "Hill" or "Mountains" or "Range", but no purpose would be achieved in thus translating *Fatshan* 佛山 which is a town in Kwangtung remote from the high country. The Mandarin word *Hua-pei ping-yuan* 華北平原 can be advantageously translated as "North China Plain", but the process is not wholly reversible. A Cantonese, while using the same written characters, would speak of the "North China Plain" as *Wah-pak ping-yuen*.

In view of the difficulties experienced in translating the meaning of a name, a more common procedure is to "romanise" the name by producing a word in alphabetic script which sounds like the spoken Chinese word. Even this method has some disadvantages. In some cases it results in a place having two names depending on the dialects of the translators.

Duplication of names is very common in China and has been brought about in several ways. Formerly, the complexities of government resulted in many places having a "county" or *Hsien* 縣 name, a "District" or *Chow* 州 name and a "prefecture" or *foo* 府 name. These still exist in such names as *Foochow* 福州, *Wenchow* 溫州 and *Wanhsien* 萬縣 although the *chow* and *hsien* may no longer denote "administrative centres." Furthermore, the names of the provinces, the major cities and principal rivers have been renamed time after time during successive dynasties. Many new names were introduced with the establishment of the 1912 Republic, and another major series of changes followed the establishment of the People's Republic, but the old names have not necessarily been forgotten. This procedure is not confined to China, and recent years have seen many changes of place-names in Korea and Taiwan and in the northeastern provinces which were once called Manchuria.

The main shortcoming of "romanisation" is that the meaning of a spoken Chinese word is conveyed by various tones and inflexions and by fine variations of sound beyond the range of a 26 letter alphabet. This results in many cases when two or more Chinese written characters with quite different meanings are spelt in the same way when "romanised." Take, for example, the word *Kiang* 江 which once referred to the Yangtze, but is now generally taken to mean a "river." It is used in *Si Kiang* 西江 ("West River"), in *Kiangsu* 江蘇 ("*Su* River Province") and *Kiangsi* 江西 ("West of the Kiangsu River Province" according to SPENCER, 1941). It retains its meaning in *Heilungkiang* 黑龍江 but the *kiang* 疆 in *Sinkiang* 新疆 ("Western Territory") is the romanisation of a completely different Chinese character meaning "a boundary."

As another example, an English-language map of China shows six towns called *Kihsien*, 冀 and 蓟 in Hupeh, 吉 and 祁 in Shansi, and 杞 and 浚 in Honan. In Chinese these are not only represented by different characters but are pronounced differently.

The directions *si* ("west"), *pei* ("north"), *tung* ("east") and *nan* ("south") enter into many place-names. They are not normally translated, though their meanings are quite important. Thus, *Shansi* 山西 means "West of the Taihang Shan" which are the ranges of the Shansi–Hopeh border. *Kwangsi* 廣西 is "West Kwang" to distinguish it from *Kwangtung* 廣東 or "East Kwang," *Kiangsi* 江西 is "west of Kiangsu," while *Shensi* 陝西 also must have once meant something similar. *Pei* 北 is found in the ancient name *Peking* 北京 ("Northern Capital") which has been revived in place of *Peiping* 北平 ("Northern Peace") in recent years, and in *Hopei* 河北 ("North of the Hwang Ho") and *Hupeh* 湖北 ("North of the Lake known as Tung Ting Hu"). *Tung* appears in *Shantung* 山東 ("Province East of the Mountains"). *Nan* 南 is found in *Honan* 河南 and *Hunan* 湖南 which are "South of the River" and "South of the Lake," in *Yunnan* 雲南 as "South of the Clouds" and *Nanking* 南京 as "Southern Capital".

There are several recognised methods of "Romanising" Chinese names. Herein, place-names are spelt as in the Chinese Post Office forms which are also used in BARTHOLO-MEW's Times Atlas (1958). In a few cases, second or third alternative names have been given to assist identification. The locations of nearly all places referred to in the text are shown in Fig.1, 2, 27, 29 and 30, but a few others may be identified by the latitude–longitude positions which are given as footnotes.

Climatological literature

There are two distinct periods in the literature dealing with the climates of China and neighbouring countries. The first or "classical" period started in the 3rd century B.C. or earlier. Here the agricultural bias was paramount, and interest was focused on seasonal changes in the fertile and populated river valleys. The "Agricultural Calendar" which has served Central China faithfully for two thousand years is still in general use, though it has only limited application in other parts of the country.

The modern instrumental period in the meteorological literature commenced before the close of the 19th century, and sufficient formal observations had been accumulated by the 1930's to encourage statistical analyses. However, the bulk of the data still referred to the most densely populated regions, and most unfortunately the observations covered inadequate and different periods of time.

The climatological observing network has greatly expanded in recent years, and CHENG KUO (1963) estimates that the number of weather observations and meteorological stations in China increased 30-fold during the period 1949–1963. Therefore, the most recent studies are based on an excellent coverage of observing stations, but in much of the region long-period data are still lacking.

The classical period

Climatological studies dating back as far as two thousand years are recorded in such classics as the *Book of the Chou Dynasty*, *Lu's Spring and Autumn Annals*, the *Book of Huai Nan Tzu* and the *Records of Rites*.[1] These works were mainly based on phenological observations. They refer to the low-lying country along the Wei Ho and the Hwang Ho (Yellow River) and to the river valleys north of the Chinling Hills in the provinces of Shensi and Honan.

Lu's Spring and Autumn Annals are of particular importance because they contain an account of the "Agricultural Calendar" which is still in use today with little change (Table I). The "Agricultural Calendar" divides the Chinese year of twelve lunar months into 24 fortnightly periods or "chi". These periods correspond to the equinoxes and

---

[1] Chi Chung Chou Shu (245–239 B.C.), Lu Shih Chun Chiu (239 B.C.), Huai Nan Tzu (120 B.C.) and Li Chi (70–50 B.C.) respectively.

TABLE I

THE CHINESE AGRICULTURAL CALENDER

	Chinese name	Translation	Beginning
1	Li Chun	Beginning of spring	4 or 5 February
2	Yu Shui	The rains	19 or 20 February
3	Ching Che (or Chih)	Awakening of creatures (from hibernation)	6 or 7 March
4	Chun Fen	Spring equinox	21 or 22 March
5	Ching Ming	Clear and bright	5 or 6 April
6	Ku Yu	Grain rain	20 or 21 April
7	Li Hsia	Beginning of summer	6 or 7 May
8	Hsiao Man	Less fullness (of grain)	21 or 22 May
9	Mang Chung	Grain in ear	6 or 7 June
10	Hsia Chih	Summer solstice	21 or 22 June
11	Hsiao Shu	Lesser heat	7 or 8 July
12	Ta Shu	Greater heat	23 or 24 July
13	Li Chiu	Beginning of autumn	8 or 9 August
14	Chu Shu	End of heat	23 or 24 August
15	Pai Lu	White dews	8 or 9 September
16	Chiu Fen	Autumn equinox	23 or 24 September
17	Han Lu	Cold dews	8 or 9 October
18	Shuang Chiang	Descent of hoar frost	24 October
19	Li Tung	Beginning of winter	7 or 8 November
20	Hsiao Hsueh	Lesser snow	22 or 23 November
21	Ta Hsueh	Greater snow	7 or 8 December
22	Tung Chih	Winter solstice	23 or 24 December
23	Hsiao Han	Lesser cold	5 or 6 January
24	Ta Han	Greater cold	20 or 21 January

solstices, to distinctive climatic periods and to important phases of plant growth. The calendar has limited application outside Shensi and Honan because China contains so many different rainfall regimes and because the times of maximum and minimum temperature differ from one place to another by two weeks or more (P. K. CHANG, 1934). The first descriptions of a "monsoon climate" over central and southern China appeared in *The Meaning of Popular Traditions and Customs* and *The Records of Airs and Places*[1] which were written in the Han Dynasty of the 2nd and 3rd centuries A.D. Various weather-diaries were kept during succeeding years, and in the 19th century BIOT (1840) and others made systematic surveys of climatic changes from the indications in ancient records.

NEEDHAM and WANG LING (1959) believe that thermometers were introduced in China about 1670 A.D. and rain-gauges at some time prior to 1770, though COOK (1964) considers that the rain-gauge was invented in Korea about 1441. The results of several studies using instrumental records were published towards the end of the 19th century, some of the best known being by BLANFORD (1876), FRITSCHE (1878), CHEVALIER (1895) and FROC (1900, 1901).

---

[1] Feng Su Thung I (175 A.D.) and Feng Thu Chi (3rd century).

4

The modern period

More recent publications range from summaries of the observations of a single station to large-scale studies of East Asia. In the first category, GHERZI's (1951) summary of observations at Shanghai (Zikawei) used data covering the 65 years, 1873–1937, and the Royal Observatory (ANONYMOUS, 1963) summarised up to 72 years of observations at Hong Kong during 1884–1939 and 1947–1962. However, these long-period series are unusual. There are few published records of continuous observations covering periods of more than 20 years, and the average period of observation for 215 stations listed by GHERZI (1928, 1951) was only 10 years. Typical of the large-scale studies are GHERZI's (1928) rainfall maps and the climatological charts of the Central Weather Bureau of China (ANONYMOUS, 1953, 1955).

Climatic controls

Topography

Some of the important factors affecting the climates of the region are topography, the large latitudinal extent of China, coastal currents and the monsoon winds and major synoptic patterns. Nearly all studies of the climatology of Southeast Asia stress the influence of topography. LU (1937) describes the large part played by the mountain ranges and passes in diverting and blocking the waves of cold air from Siberia and Mongolia which flow southwards over China during the winter. P. K. CHANG (1941) describes the climates of the province of Szechuan with reference to topography, while KENDREW (1941) compares winter conditions in the sheltered valleys of the Yangtze with those along the exposed coasts.

In the centre of the continent of Asia there are more than two and a half million km² of land above 3,000 m M.S.L. Much of the area is above 5,000 m and there are many higher peaks, including Chomolangma (Mount Everest) which rises to 8,848 m. The high country, which is generally known as the Tibetan Plateau, extends beyond the terri- torial boundaries of Tibet to cover most of the provinces of Chinghai and Changtu[1] and the western parts of Kansu and Szechuan (see Fig.1–2). The plateau, by blocking the entry of low-level westerlies and southwesterlies, has a great effect on the general circulation and on weather conditions in China (KOO, 1951; YEH, 1955).

North of the Tibetan Plateau the ground falls away fairly rapidly, and beyond 38°N latitude the general ground level is between 1,000 and 2,000 m M.S.L. However, the Gobi Desert is at a still lower level and the Tarim and Dzungaria Basins descend to 784 and 191 m respectively, at their lowest points. There are two major mountain ranges in the northern part of this region with tops mainly between 3,000 and 5,000 m: the Altai Mountains lie along the western fringe of the Mongolian People's Republic and the Tien Shan Range runs east–west across the middle of Sinkiang Province.

East of the Tibetan Plateau the ground-level falls away to below 200 m over large areas

---

[1] In some recent publications the boundary between Changtu (Chamdo) and Tibet is omitted.

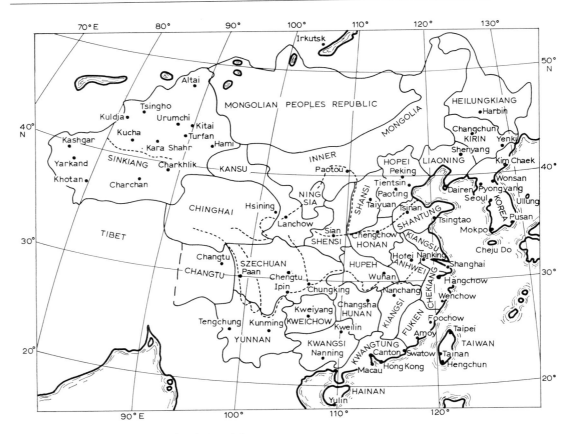

Fig.1. Map of Central and East Asia.

in a coastal strip 1,000 km wide. The low land lies mainly east of a line drawn from Harbin to the Gulf of Tongking and includes the North China Plain and the Han Basin. The most prominent features are the highlands of North Korea, the Chinling Hills which protrude eastwards into the North China Plain along 34°N latitude, and the Nanling Hills which cross the country at 25°N with an eastward extension through Fukien and Chekiang. InSouth Korea the level of the country varies from about 100 m M.S.L. in the west to over 500 m in the east.

Many writers have stressed the importance of the Chinling and Nanling Ranges. According to YAO (1946), winter and spring rains in the middle Yangtze are largely due to wave disturbances on a quasi-stationary front which forms where the Nanling Hills block the southward advance of cold outbursts. CHU and YUAN (1963) consider that the Chinling Hills mark the boundary between the temperate and subtropical climate zones because, although bamboo and tea and mandarin oranges thrive south of the hills, they survive only in very sheltered localities in the north. The Chinling Hills do not in fact extend completely across the lowlands. They are parallel to the mean isotherms and isohyets, and the many literary references to them as climatic boundaries may be more a matter of convenience than an indication of an actual physical barrier. Chu and Yuan consider that the Nanling Range marks the boundary between the subtropical and the tropical zones, because south of the range the climate is more adequately described by the seasonal variation of rainfall than by that of temperature.

In Taiwan, the western lowlands cover a quarter of the total area of the island at an altitude of less than 100 M.S.L. However, greatly differing rainfall regimes are determined by the steep central and eastern mountains which are mainly above 1,500 m with many peaks over 3,500 m.

Around the coastal strip of Hainan Island and for about 30 km inland the general ground level is less than 200 m, but it rises to over 1,500 m in the centre of the island. Compared with other parts of China, the climate of Hainan is fairly uniform.

Although some degree of homogeneity of climate is found over quite large areas of generally similar topography, there is a danger in attempting to oversimplify the picture. The most fertile valleys and the centres of greatest population in China are in comparatively small areas of very distinctive climate. The climates of the sheltered valleys of North Taiwan, the Han Basin of Hupeh and the Red Basin of Szechuan differ greatly from conditions in the surrounding areas.

Continentality and latitude

The distance from the western boundary of Sinkiang to the east coast of Korea is more than 4,800 km. The nearest coastline to western China is 1,500 km away across the Himalayas to the Arabian Sea, and the climates of western and central China are distinguished by temperature extremes greatly exceeding those recorded in other parts of the country.

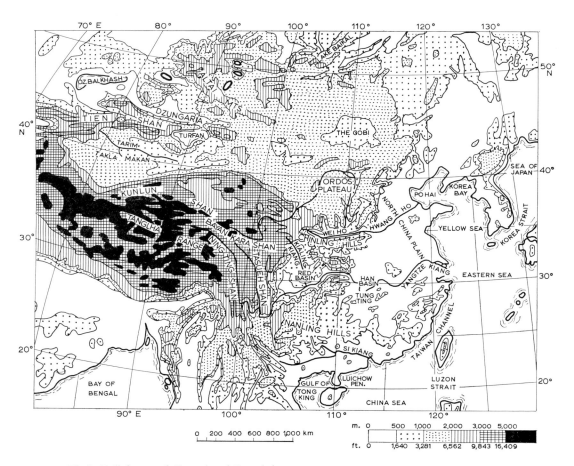

Fig.2. Relief map of Central and East Asia.

Thus, the average diurnal variation of temperature at Hami in Sinkiang is 14°C, while mean monthly temperatures range from −12°C in January to 30°C in July. At Yentai[1] in Shantung, the mean diurnal range is only 6° to 7°C, while mean monthly temperatures range from −1°C in January to 26°C in July.

Sinkiang and the Inner Mongolian Autonomous Region are exceedingly dry. They are beyond the reach of the summer monsoon winds from the China Sea and the Pacific, and they are sheltered from moist southwesterlies by the Tibetan Plateau.

The large range of latitude covered by the region also contributes to the diversity of climate. Continental China stretches over more than 35° of latitude from the northern-most boundary of Heilungkiang at 53°26′ to the southernmost tip of Hainan Island at 18°14′N. January mean temperatures range from −30°C in the extreme north to higher than 18°C in the south. The differences, however, are less marked in summer with July temperatures varying only from 20°C in the north to 30°C in the lowlands of latitude 30°N.

Coastal currents

In summer the warm Northern Equatorial Current, which crosses the Pacific Ocean in the trade-wind belt, is diverted northeastwards as the Kuroshio Current about the latitude of Taiwan. This is shown in Fig.3 (ANONYMOUS, 1945, 1962b). Nearer the Asian coast, a warm southwesterly current from the tropical Java Sea covers the China and Eastern Seas and extends northward beyond Korea as the Tsushima Current. Sea surface temperatures along the Asian coast are generally high in summer ranging from 29°C around Hainan to 20°C near North Korea; but they are still one or two degrees lower than the air temperature along the entire coast (ANONYMOUS, 1961).

In winter, part of the warm Northern Equatorial Current continues to be diverted east of Japan as the Kuroshio. Some part is also diverted southward into the South China Sea, but this has no effect on the climate of the mainland north of Hainan. The main flow along the China coast is a cold current from north to south under the influence of the Northeast Monsoon winds (Fig.3). Sea temperatures along the coast fall rapidly in October. By January they reach 0°C in Po Hai (Chihli) where broken ice frequently extends 100–150 km offshore. Ice occasionally extends into Korea Bay, and further south sea-temperatures in January range from 10°C in the Yellow Sea to 20°C around Hainan. During winter the coastal waters are a few degrees warmer than the air above them (ANONYMOUS, 1961), but winter temperatures on the coast differ little from those over the low-lying country further inland.

In late winter and early spring, small anticyclonic circulations sometimes move east-ward from the continent across the Yellow Sea. On these occasions, warm moist air with a long track over tropical waters passes westwards across Luzon Strait and the Philippines into the China Sea. Sea temperatures in spring are much the same as those shown for January in Fig.3. As the moist air crosses the cool waters near the China coast, it is often cooled sufficiently to form fog or low stratus. This is known as "crachin", and the "crachin season" starts about the time of the annual temperature minimum and persists into mid-April (BRUZON and CARTON, 1930). HUNG (1951) estimates that fog

---

[1] Yentai (Chefoo): 37°30′N 121°22′E.

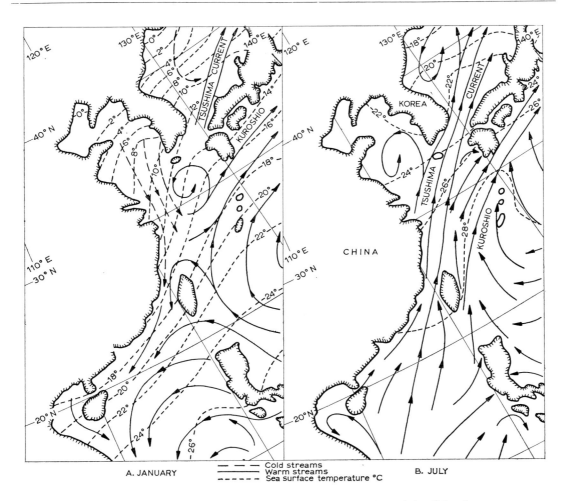

Fig.3. Stream drift and sea temperatures in the northwest Pacific and the China Seas.

occurs at Waglan Island[1] about 10 days per month during March and April, 6 in February and much less frequently in other months. The seaward limit of the crachin is nearly always sharply defined (ANONYMOUS, 1937). RAMAGE (1954) has suggested that it is usually confined to within 160 km of the China coast while S. Cheng (analyses of ships' reports, personal communication, 1964) reports that 100 km from the coast is a common limit.

Synoptic weather patterns and the monsoons

Typical weather patterns over East Asia have been described by Tu (1938), Y. S. KAO (1948), TAO (1948), C. S. CHENG (1949), THOMPSON (1951) and others, and each of these writers emphasizes the marked differences between winter and summer regimes. However, the length of each season varies greatly from place to place. P. K. CHANG (1934), taking 5-day mean temperature limits of 10°C and 22°C to define spring and autumn, finds that while winter does not exist in the southern provinces, it lasts for half the year in Sinkiang and Heilungkiang (see Table II).

---

[1] Waglan Island (6 km southeast of Hong Kong): 22°01′N 114°18′E.

TABLE II

DURATION (IN MONTHS) OF THE SEASONS (AFTER P. K. CHANG, 1934)

Region	Spring	Summer	Autumn	Winter	Spring and autumn
Sinkiang	2–3	2	2	5–6	—
Northeast China	2–3	1½–2½	2	5½–7	—
Inner Mongolia	2–3	1–3	1½–2½	5½–6½	—
North China	1½–3	2–4½	1½–2	4½–6	—
Central China and the Yangtze	2–3	3–5	2–3	2½–4½	—
Yunnan-Kweichow Plateau	—	—	—	2–3	9–10
South China	—	6–8	—	—	4–6

Winter

From October onwards, there is a rapid fall of temperature in the centre of the Asian continent and a cold anticyclone forms over Siberia and Mongolia. At Irkutsk, mean monthly temperatures fall from 8°C in September to 1°C in October and to about −20°C in December and January. The mean monthly pressure at Irkutsk is between 1,005 and 1,015 mbar throughout the summer, but rises to 1,023 mbar in October and to over 1,032 mbar in mid-winter.

The continental anticyclone gradually spreads out until by January its influence is felt over the whole continent. Associated with the anticyclone, northeasterlies develop over China and persist throughout the winter. The season is known as the "northeast monsoon" or "winter monsoon". The air over the region is cold, dry, stable polar continental air (TU, 1938), and the monsoon flow is gradually established in successive bursts of cold northeasterlies commencing in October and progressing further and further southward to cover all China before the end of November. Since mean monthly pressures over southern Kwangtung only rise to about 1,020 mbar in winter, a steep north–south gradient of pressure persists over the whole of East Asia.

The frequent southward-moving outbursts of cold air which occur throughout the winter are variously called "cold waves" or "surges of the northeast monsoon". LU (1937) considers that, although some of the cold waves crossing China may come from the arctic, most of them originate in Mongolia. Cold air from the arctic passing on to the continent near Novaya Zemlya can spread without hindrance across the western Siberian plain where none of the land is higher than 200 m. To the south of this plain, however, Tien Shan and the Altai Mountains block the advance of the cold air, except for a few very vigorous outbursts which may be channelled through the Dzungaria Basin to reach as far as southeastern Sinkiang. Cold waves from the arctic produce winter rains in a few exposed parts of Sinkiang (LU, 1937) and contribute to an annual rainfall of 600–700 mm in Tien Shan.

The winter anticyclone is nearly always centred over Mongolia (LAHEY et al., 1958) where the general ground level is between 1,000 and 2,000 m M.S.L. Mean winter temperatures in that area are below −10°C and probably often below −20°C; average temperatures at Paotou during December and January are −10°C and −12°C respectively.

As air outflowing from the continental anticyclone spreads down to Hopei, Liaoning and the North China Plain, the successive cold fronts sweep southeastward across China and Korea at intervals of about one week throughout the winter. The advance of a cold front over southern China is often associated with the development of a wave depression over the Yangtze area. According to Lu (1937), the cold fronts extend only 1,500–2,000 m above ground level, so that their progress is greatly affected by topography. The Chinling and Nanling Hills frequently delay their advance, and in many places progress of the cold air is restricted to narrow valleys. Thus the Han Valley and Tung Ting Basin permit the passage of cold waves down into the valleys of Hunan and Kwangsi, but the Red Basin of Szechuan and the deep valleys of Yunnan are affected only on the rare occasions when particularly vigorous waves penetrate the Kialing Kiang Valley. The southeastern provinces are sheltered to some extent by the Nanling Hills, but they are quite exposed to any cold outbursts which travel along the coast. These are the occasions when cold air from Mongolia has passed freely across Hopei, Shantung and the Yellow Sea to be funnelled southwestwards through the Taiwan Channel to the southeastern provinces of Fukien and Kwangtung.

A cold front is frequently accompanied by gales which persist for days in exposed places, particularly over the North China Plain, the lower Yangtze Valley and the Taiwan Channel. With the frontal passage, winds change from southerlies or easterlies to northerlies or northeasterlies, humidity becomes very low and temperature drops by 10° or even 20°C. Lu (1937) states that the absolute minimum temperature of −14°C at Nanking followed a frontal passage.

Apart from the special case of Sinkiang described earlier, cold fronts over northern China are rarely accompanied by rain. In the central provinces winter rains are mainly frontal and are frequently associated with wave-depressions forming on the front and moving eastwards. In the southern provinces most of the meagre winter rainfall can be attributed to cold fronts, particularly when in association with high-level disturbances·

Spring

In March and April the continental anticyclone gradually weakens. Cold-surges from the north become indistinct and less frequent, and gradually increasing incursions o warm, moist tropical air from between south and east replace the cool northeasterlies. Since coastal waters are still comparatively cold (Fig.3), the outbreaks of warm air produce widespread stratus, drizzle and fog which may persist along the coast for days In the south the fogs are most frequent in March, but the maximum is delayed till June about 30°N and to July in the Yellow Sea. As sea-temperatures rise and atmospheric instability increases with the advance of summer, the fogs become less frequent.

Summer

In summer a large semi-permanent thermal depression develops over the southwestern provinces of China where pressure is about 5 mbar less than that of the eastern provinces (GHERZI, 1944). According to Y. S. KAO (1948), and THOMPSON (1951), this results in a general southwesterly flow of air over China at levels above 600 m except over the Mongolian Plateau where northwesterlies intrude (Fig.4). The southwesterlies, which have

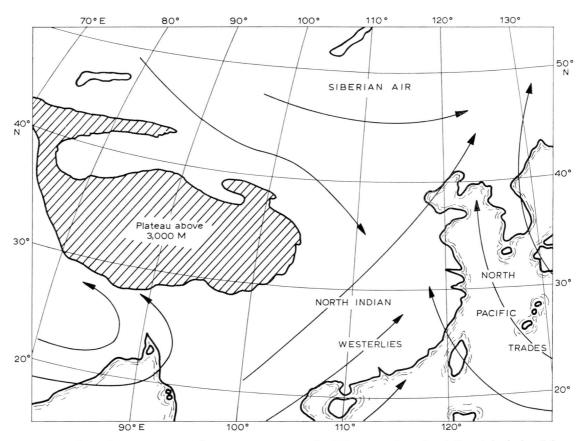

Fig.4. The most common flow patterns at 600 and 3,000 m over Central and East Asia during July. (After THOMPSON, 1951.)

passed over the Indian Ocean and South China Sea, are warm, moist and conditionally unstable (TU, 1938). Consequently, there is considerable convective activity overland, and scattered showers and thunderstorms contribute largely to the summer rainfall maximum which is observed over nearly all the region.

The season is generally called the "summer monsoon" but sometimes the "southeast monsoon" (CHU, 1934), because at ground-level southeasterlies occur more frequently than southwesterlies in most places. The development of the monsoon is a gradual

TABLE III

PERCENTAGE DISTRIBUTION OF SURFACE WINDS IN JULY AND AUGUST (KOREA 1957–1959; CHINA 1956–1995)

Place	Wind direction								
	N	NE	E	SE	S	SW	W	NW	calm
Wonsan	3	17	18	1	2	15	6	2	36
Pyongyang	4	11	21	10	8	7	7	12	20
Shanghai	3	8	17	23	17	11	8	5	8
Foochow	9	8	7	37	9	3	3	7	17
Swatow	4	6	17	16	19	15	10	3	10
Canton	3	9	21	26	16	5	4	1	15

TABLE IV

AVERAGE NUMBER OF TROPICAL CYCLONES WITHIN 5°N–30°N AND 105°E–150°E DURING 70 YEARS 1884–1953

Month	Average number
January	0.3
February	0.1
March	0.1
April	0.3
May	1.0
June	1.5
July	3.8
August	4.4
September	4.4
October	3.0
November	2.1
December	0.9
Year	21.9

process covering about 2 months and there is no clearly defined advance such as occurs in India (Tu and HWANG, 1945). CHU (1934) states that the monsoon starts on the southern coast of China in April but does not reach the northern provinces until the end of June or the beginning of July. The pattern of winds at low levels in summer is greatly confused by topography. According to TAO (1948), southeasterlies cover most of the region from June to August and northeasterlies in September, and some evidence of the summer southeasterlies can be seen in Table III. In southern China, summer southwesterlies or southerlies alternate with incursions of trade-wind easterlies from the North Pacific (See Fig.4 and Swatow in Table III) and each change is usually accompanied by convergence and precipitation.

Nearly every summer, some parts of East Asia are affected by tropical cyclones of varying intensity, ranging from weak depressions to mature typhoons[1] containing hurricane force winds. Statistical analyses of past storms have been published by FROC (1920), STARBUCK (1951) and CHIN (1958). Table IV shows about 22 cyclones per year in the northwestern Pacific and China Sea, most of them occurring between July and October. However, recent improvements in the observational network and techniques are resulting in the detection of more disturbances than previously. During the years 1947–1964, the annual average was 31 tropical cyclones with a maximum of 45 in 1961 and a minimum of 20 in 1951.

Tropical cyclones frequently pass over the coasts of China and Korea during July to September, while the southern provinces are also sometimes affected as early as May and as late as mid-November. From mid-November to April very few cyclones reach the mainland.

During the worst months (July to September), most of the cyclones originate over the warm North Equatorial Current between the Marianas and Luzon and commence moving towards the west–northwest. Approximately 50% of them continue in this direction until

---

[1] The Chinese "taai fung" typhoon means "big wind".

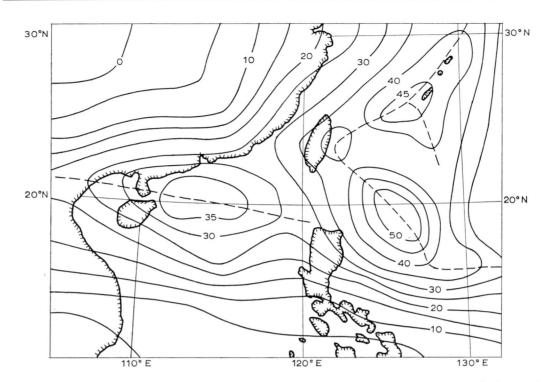

Fig.5. Number of tropical cyclones passing through each square of 2½° latitude and longitude over the northwest Pacific and the China Seas during August 1884–1953. Broken lines are axes of maxima. (After CHIN, 1958.)

they strike the south China coast, but the remainder recurve northwards between 120°–130°E longitude in the direction of Korea and Japan. This is illustrated in Fig.5 (CHIN, 1958), which shows the most common cyclone tracks for August during the years 1884–1953.

An analysis by KAO and TSANG (1957) of the number of typhoons (presumably mature hurricanes) crossing various sections of the coast indicates that the Fukien–Taiwan sector is particularly vulnerable (see Table V).

Well known among the many detailed accounts of the incidence and effects of tropical

TABLE V

NUMBER OF TYPHOONS CROSSING DIFFERENT SECTIONS OF THE COAST DURING 1884–1955 (AFTER KAO AND TSANG, 1957)

Coastal region	Number of typhoons
Korea and further east	87
Liaoning to Shantung Peninsula	39
Shantung Peninsula to Shanghai	22
Shanghai to Wenchow	34
Wenchow to Foochow	30
Foochow to Swatow	90
Swatow to Canton	43
Canton to Hainan	93
Total	438

cyclones in East Asia are those of ALGUE (1904) and GHERZI (1939). Few years pass without reports of loss of life and tremendous damage by hurricane winds, torrential rain and flooding somewhere along the coast. In 1881, 300,000 people died at Haiphong during a typhoon (TANNEHILL, 1927); 60,000 lives were lost in Swatow in 1922 (HEYWOOD, 1950) and 11,000 in Hong Kong in September 1937 (JEFFRIES, 1938).

Tropical cyclones usually weaken as they pass overland and it can be seen from Fig.5 that few of them persist more than 500 km inland. On the coast, however, they occur with sufficient frequency to make a substantial contribution to the annual rainfall.

The rise of sea-level with a mature tropical cyclone accounts for more damage and loss of life on the China coast than the wind does. Near the centre of a typhoon the sea-level is raised in a storm surge, and this can be very dangerous when the arrival of the typhoon coincides with the normal time of high tide. The greatest storm-surge (recorded tide-height minus predicted tide-height) at Hong Kong was a little over 2 m, but surges 3 times this height have probably occurred on parts of the coast where the waters are channelled into narrowing bays and estuaries (WATTS, 1959).

Autumn

In September, temperatures decrease rapidly and showers become less frequent. Tropical cyclones become progressively less intense and fewer reach the mainland. From mid-October onwards, occasional cool northeasterly surges pass southward over the country and the winter monsoon becomes well established during November.

Wind

Winter winds

During the "winter monsoon" or "northeast monsoon", the centre of highest pressure over Mongolia is generally west of 110°E longitude. Consequently, the low-level outflow over Heilungkiang, Kirin and Korea is most often from the northwest or west. Thus, at Yenki northwesterlies or westerlies prevail for more than 40% of the time during December to February and winds from other directions for less than 20% of the time (see Table VI).

Further south there is a tendency for the stream to turn to the right under the influence of Coriolis force. Therefore, during December–February northerlies are just as common as northwesterlies on the Ordos Plateau, in Hopei and Shansi and about Po Hai, as can be seen from data for Shihkiachwang[1] and Dairen in Table VI. The swing to northerlies is discernible over nearly all this area, though the rough terrain also exercises a considerable influence on wind direction at places. Over the North China Plain and further to the south the stream turns even more to the right until northeasterlies become at least as prominent as northerlies (e.g., Nanyang[1] and Wuhan[2] in Table VI), and easterlies are very common on parts of the south China coast (e.g., Tungshan[1], Hong Kong and Yulin).

---

[1] Shihkiachwang: 38°04'N 114°28'E, Nanyang: 33°06'N 112°31'E, Tungshan: 230°45'N 117°30'E.

[2] Wuhan includes Wuchang, Hankow and Hanyang.

TABLE VI

PERCENTAGE OF TIME IN EACH SEASON WITH WIND FROM VARIOUS DIRECTIONS[1]

	Season	Wind direction								
		N	NE	E	SE	S	SW	W	NW	calm
Yenki	Dec.–Feb.	1	2	3	2	2	8	30	12	40
	Mar.–May	3	12	9	2	2	7	31	10	24
	June–Aug.	4	26	26	3	1	4	10	4	22
	Sep.–Nov.	3	6	7	2	2	8	25	9	38
Shihkiachwang	Dec.–Feb.	19	8	6	12	7	2	5	12	29
	Mar.–May	13	9	6	19	12	3	7	12	19
	June–Aug.	15	10	6	15	8	2	4	11	29
	Sep.–Nov.	16	7	4	12	8	2	6	14	31
Dairen	Dec.–Feb.	34	7	6	5	6	7	9	21	5
	Mar.–May	15	3	10	23	11	9	9	18	2
	June–Aug.	9	5	16	31	12	5	6	11	5
	Sep.–Nov.	25	4	3	10	18	14	9	13	4
Nanyang	Dec.–Feb.	9	29	12	3	7	10	4	3	23
	Mar.–May	9	26	11	6	9	12	6	2	19
	June–Aug.	10	25	11	8	13	9	3	2	19
	Sep.–Nov.	14	27	10	3	6	8	4	2	26
Wuhan	Dec.–Feb.	22	22	12	10	3	5	4	6	16
	Mar.–May	18	17	16	18	4	6	3	6	12
	June–Aug.	11	14	13	17	11	14	5	4	11
	Sep.–Nov.	22	24	15	9	2	3	3	8	14
Tungshan	Dec.–Feb.	12	63	17	1	2	1	1	2	1
	Mar.–May	8	40	16	5	14	10	1	1	5
	June–Aug.	3	15	11	8	25	25	5	1	7
	Sep.–Nov.	16	53	17	2	3	4	2	2	1
Hong Kong	Dec.–Feb.	15	20	40	9	1	2	3	5	5
	Mar.–May	5	13	46	12	7	5	4	2	6
	June–Aug.	1	6	23	12	16	18	11	4	9
	Sep.–Nov.	15	19	36	10	3	2	4	4	7
Yulin	Dec.–Feb.	1	42	28	6	4	1	0	1	17
	Mar.–May	1	21	30	19	12	3	1	1	12
	June–Aug.	1	11	11	12	13	10	14	4	24
	Sep.–Nov.	3	36	18	3	5	4	3	3	25

[1] Based on hourly observations from 1884–1939 and 1947–1960 for Hong Kong, and observation at 00, 06, 12 and 18 G.M.T. during August 1956–October 1959 for other places.

In the mountains and valleys of the western provinces, wind-directions vary greatly in winter and there is no clear overall pattern.

Table VII shows the percentage frequency of strong winds recorded at different locations. The strongest winds of winter are recorded on the exposed parts of the east coast.

On the coast, the winter monsoon wind is mainly moderate, but it temporarily increases with each outburst of cold air and thereafter may remain strong for several consecutive days. Winter monsoon outbursts are characterised by violent short-lived snow-storms in northern coastal areas, by gales lasting a week or more in the narrow Taiwan Channel and by very high seas over the China Sea. Inland, gales are generally less frequent and less persistent.

TABLE VII

PERCENTAGE OF 6-HOURLY OBSERVATIONS IN EACH SEASON WHEN WIND IS WITHIN HIGHER RANGES OF SPEED[1]

Place	Season	Speed (m/sec)		
		11–14	15–21	>21
Yenki	Dec.–Feb.	1.2	—	—
	Mar.–May	2.0	0.2	—
	Jun.–Aug.	—	—	—
	Sep.–Nov.	1.3	0.1	—
Shihkiachwang	Dec.–Feb.	0.4	—	—
	Mar.–May	0.2	—	—
	Jun.–Aug.	—	—	—
	Sep.–Nov.	—	—	0.1
Dairen	Dec.–Feb.	5.7	2.0	—
	Mar.–May	4.3	1.2	0.1
	Jun.–Aug.	1.0	0.2	—
	Sep.–Nov.	3.9	0.4	0.1
Nanyang	Dec.–Feb.	0.2	—	—
	Mar.–May	0.5	—	—
	Jun.–Aug.	—	—	—
	Sep.–Nov.	0.1	0.1	—
Wuhan	Dec.–Feb.	0.1	—	—
	Mar.–May	0.4	—	—
	Jun.–Aug.	0.2	—	0.1
	Sep.–Nov.	0.1	0.1	—
Tungshan	Dec.–Feb.	25.6	11.3	0.4
	Mar.–May	12.3	7.8	0.5
	Jun.–Aug.	2.8	0.7	0.1
	Sep.–Nov.	22.2	8.0	1.0
Hong Kong	Dec.–Feb.	2.0	0.1	—
	Mar.–May	2.0	0.2	—
	Jun.–Aug.	1.0	—	—
	Sep.–Nov.	0.4	0.1	0.1
Yulin	Dec.–Feb.	—	—	—
	Mar.–May	—	—	—
	Jun.–Aug.	—	0.1	0.1
	Sep.–Nov.	0.1	—	—

[1] Based on observations at 00, 06, 12 and 18 G.M.T. during August 1956–October 1959.

Tables VI and VII show that the strong winter winds from a northerly quarter persist into March–May; but during March–May there is also some diminution of wind-strength and the gradual replacement of northerlies by winds from between east and south.

Summer winds

From May till September, winds are generally light, apart from brief squalls associated with thundershowers and the diurnal freshening of the sea-breeze along the coast. However, summer is also the typhoon season and very few summers pass without some section of the coast being devastated by hurricane-force winds. July to September is the most dangerous season for both China and Korea, but gales have been recorded in the southern provinces and Taiwan as early as May and as late as November. Since a typhoon weakens as it passes inland, hurricane-force winds are rarely observed more than 200 km inland and gale-force is rarely maintained more than 500 km in from the coast.

The counterclockwise circulation of a typhoon is usually several hundred km in diameter. As most typhoons are moving in a west–northwesterly direction when they strike the coast, the strongest winds are usually from the northeast in the early stages and from the southeast or south later. Most typhoons approach the coast at a speed of about 5 m/sec but those which recurve in the eastern sea usually accelerate to more than 10 m/sec.

Typhoon winds have little influence on the long-period wind averages because typhoons rarely affect any particular location more than once during each season and because winds of hurricane force (33 m/sec) are rarely maintained at any one place for as long as 2 h. Gale force winds (17 m/sec) occur in Hong Kong on an average of only once per year (HEYWOOD, 1950), with many gale-free years and a maximum of five in 1964. The average duration of gale-force winds is only about 7 h with a maximum of 27 h (see Table VIII).

TABLE VIII

DURATION OF TYPHOON GALES AT HONG KONG IN 72 YEARS, 1884–1939, 1947–1962 (ANONYMOUS, 1963)

Duration (h)	Number of occasions
1–5	44
6–10	21
11–15	16
16–20	4
21–25	1
>25	1
Total	87

TABLE IX

NUMBER OF OCCASIONS OF VARIOUS VALUES OF MAXIMUM HOURLY WIND IN ALL TYPHOONS WHICH PRODUCED GALES AT HONG KONG DURING 1884–1941, 1946–1962

Maximum hourly wind (m/sec)	Number of occasions
17–19	27
20–22	26
23–15	13
26–28	10
29–31	6
32–34	2
35–37	2
Total	86

TABLE X

RATIOS BETWEEN THE HOURLY MEAN WIND AND THE MEAN WIND OVER SPECIFIED SHORTER PERIODS AT HONG KONG

Duration	Ratio
1 h	1.0
10 min	1.05
1 min	1.28
6 sec	1.70
Gust on Dines anenometer	2.05

TABLE XI

FATALITIES AND DAMAGE WITH TYPHOONS IN HONG KONG

Year	Ocean-going ships sunk or ashore	Junks sunk	Junks damaged	Dead or missing	Injured
1937	28	1,800		11,000	—
1960 (Mary)	6	352	462	56	127
1962 (Wanda)	36	1297	756	183	—
1964 (Ruby)	20	32	282	44	300

K. T. C. CHENG (1960) has published the absolute maximum "instantaneous" winds recorded at twelve stations in and near Taiwan. These included 65.7 m/sec at Lan Yu, 52.0 at Pengchia Yu[1], 49.7 at Tung Shan (Yu Shan) and 45.0 at Hualien (see Fig.29). According to GHERZI (1951), wind-speeds (presumably of short duration and measured by various means) have attained 78 m/sec at Santuao[1], 72 m/sec near Wenchow and 70 m/sec at Hengchun, while the strongest gust ever recorded at Shanghai was 42 m/sec. At Hong Kong, a Dines anemometer has measured gusts of 75 m/sec in 1937 (HEYWOOD, 1950) and 79 m/sec in 1962 (ANONYMOUS, 1963). However, gustiness varies considerably according to wind-direction and topography, and the Royal Observatory in Hong Kong classifies tropical cyclones according to the greatest hourly mean winds (ANONYMOUS, 1963; see Table IX).

Mean hourly typhoon winds for other places are not available for comparison and most of the observations which now enter into international exchange are either 10-min averages as prescribed by International Civil Aviation Organization or are shorter period estimates made by reconnaissance aircraft. The latter are based on the appearance of the sea and, according to JORDAN and FORTNER (1960), lie between the peak gusts and the 1-min averages. FABER and BELL (1963) have determined the ratios between the hourly mean wind and the mean wind over specified shorter periods at Hong Kong as shown in Table X. During an intense typhoon over the China Sea in September 1964, a reconnais-

[1] Pengchia Yu: 25°38′N 122°04′E, Santuao: 26°38′N 119°42′E.

sance aircraft reported short period winds of 103 m/sec. If the Faber-Bell ratios are of general application, this would be equivalent to a mean hourly speed of 50–80 m/sec which probably represents the worst conditions ever experienced in the region.

Ships and small craft suffer most from typhoons, and the effects of the three worst typhoons to strike Hong Kong in recent years are detailed in Table XI. Storm-surges rather than wind accounted for most of the fatalities in 1937 and 1962. These surges frequently accompany typhoons and were responsible for the great loss of life in the historic Haiphong and Swatow typhoons (see p.15).

A basic requirement in building design is an estimate of the velocity and frequency or wave-length of the most prominent gusts or squalls during typhoons. GHERZI (1951) considers there are two prominent groups of oscillations at intervals of 1-min and 4–7 sec, respectively, while HEYWOOD (1954) suggests periods of 2 min and 10 sec. Recently FABER and BELL (1963) have made a preliminary analysis of the power spectrum in a typhoon and found that eddies occurred at all wavelengths between 250 and 6,000 m, with one peak of maximum energy at a wavelength of 1,700 m and a secondary one at 300 m. Assuming a mean wind speed of 36 m/sec, the corresponding periods would be 47 sec and 8 sec, respectively.

Improvements in the instruments used for observing the fine structure of winds and the introduction of computer methods have facilitated analyses of this sort and a more recent power spectrum for a Hong Kong typhoon showed a peak of maximum energy at 1,000 m and a secondary peak of 330 m. The corresponding periods with a mean wind of 36 m/sec would be 28 sec and 9 sec (ANONYMOUS, 1965).

Precipitation and thunderstorms

Annual rainfall

Annual rainfall is greatest in the southeastern provinces of China and decreases towards the northwest (Fig.6). Mean annual totals exceed 2,000 mm over a large area in Kwangsi and parts of coastal Kwangtung and northern Fukien, while they reach 3,000 mm in the high country of Taiwan. East of 100°E longitude, the 1,100 mm isohyet mainly lies east–west along the Yangtze Kiang, except through the Red Basin where rainfall is over 1,250 mm. In the province of Kiangsu, annual rainfall varies from 600 mm in the north to 1,100 mm in the south (HU, 1947). Further to the north, the isohyets run northeast–southwest and mean annual precipitation is less than 250 mm over much of Kansu, Chinghai and Inner Mongolia. West of 100°E longitude, the 1,100 mm isohyet lies to the southwest of the Tanglha and Ning Ting Ranges and rainfall decreases rapidly towards the north.

In the mountainous Tasueh and Ning Ting region the mean annual isohyets mask large local variations depending on topography. Most of the rainfall is associated with the moist summer southeasterlies, and is greater on the eastern slopes than to leeward or in the valleys. Thus, at Kangting (Tatsienlu), on the east of Tasueh Shan, mean annual rainfall is 1,000 mm while Paan (Batang) in the Upper Yangtze Valley has only 560 mm (HANSON-LOWE, 1941).

Little information has been published regarding conditions in northwestern China, but

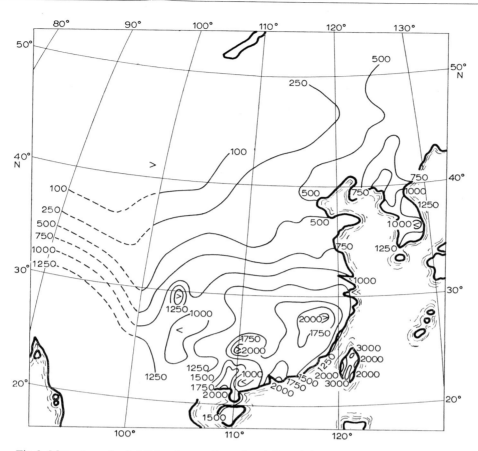

Fig.6. Mean annual rainfall (mm) over Central and East Asia.

rainfall is known to be very sparse there. Mean maps of the Central Weather Bureau (ANONYMOUS, 1953) indicate a large area with less than 100 mm annually, comprising the northernmost parts of Kansu, Chinghai and most of Sinkiang. LU (1944) confirms that annual totals of this order are found in the Gobi and the Tarim Basin (or Takla Makan) and that about 30% of the rain is brought about by local convection. C. Y. CHANG (1949) recorded an annual rainfall of 76 mm at Kucha (1928–1930), 41 mm at Khotan (1942–1943) and 5 mm at Charkhlik, but considered that total falls were 50% higher in the hills surrounding the Tarim. Mean annual rainfall exceeds 200 mm in the Kuldja Plain and other parts of Dzungaria (Table XXV) and is probably greater over some of the high country of Tien Shan and the Altai Mountains.

In Taiwan, average rainfall is a little over 1,500 mm on the northwest coast, but elsewhere mainly 2,000 mm on the lower ground and 3,000 mm or more throughout the central ranges. The smaller islands along the Chinese coast have generally about 20–40% less rainfall than places nearby on the coast (LEE, 1936). The mean rainfall of Korea is 600 mm in the northeast but otherwise mainly 1,000–2,200 mm.

The distribution of average rain-days (with 0.1 mm or more) follows much the same pattern (Fig.7). There is one area with more than 150 rain-days annually inland through Fukien and Chekiang, and another stretching from Hunan through Kweichow into the low-lying parts of Szechuan east of Tasueh Shan. Apart from these areas of maximum frequency there are mainly 100 days per year from the Yangtze southward. North of the

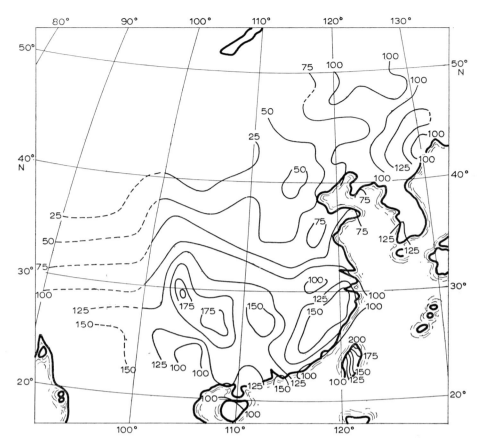

Fig.7. Mean number of days each year with precipitation (0.1 mm or more) over Central and East Asia.

Yangtze there are 50 to 70 days per year, except in Heilungkiang where there are 100 days annually and in Sinkiang where there are only about 25.

In Taiwan, rain falls most frequently on the northernmost tip with 200 days per annum and is least frequent along the western coast with 100 days. In Korea there are generally more than 100 rain-days per annum but there are more than 120 in the southwest of the country and inland in the north.

The most striking feature of the annual rainfall is its irregularity. In some years vast areas are inundated by flooded rivers while at other times there are serious drought conditions. Although Tainan has mean rainfall of 1,811 mm, annual totals have ranged from 685 mm to over 3,521 mm. Throughout the region most of the rain falls in summer, and annual totals have been small on occasions when persistence of the Siberian anticyclone as late as May has delayed the development of a moist unstable atmosphere over China. Lu (1944) considers that, although typhoons contribute only 5% of the annual rainfall inland, and 10–20% on the East Coast and in Shantung, they provide 30–35% of the total on the south China coast. Other investigators have arrived at different estimates, mainly because of the difficulty in agreeing on a definition for "typhoon" and in deciding what part of the rainfall in the wake of a typhoon is directly associated with the typhoon. Thus, K. K. Chang (1958) estimates that the average contribution of typhoons is 21% at Canton, 17% at Foochow, 11% at Shanghai, 4% at Nanking, 3% at Peking and only 2% at Wuhan.

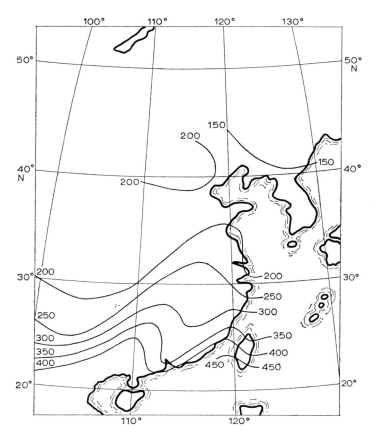

Fig.8. Standard deviation of annual total rainfall (mm).

During 33 years (1921–1939, 1947–1960), rainfall associated with tropical cyclones constituted 24% of the total in Hong Kong, while during the months of July, September and October it amounted to 40% of the total. (In this estimate, typhoon rainfall was considered to commence on the day the cyclone's centre first came within 500 km of Hongkong and to finish 3 days after.) On an average, three typhoons strike the China coast between Shanghai and Canton each year (KAO and TSANG, 1957); but there are many records of sub-normal rainfall on this coast in years when successive storms have either passed along 20°N latitude to the south of Kwangtung or recurved towards Japan without entering the China Sea (see Fig.5). However, not all the fluctuations in annual rainfall can be ascribed to unusual local synoptic patterns or sequences. A recent analysis of Hong Kong records covering 110 years showed that 1859, 1895, 1929 and 1963 were very dry years, thus indicating the influence of a 35–year Brückner cycle (BELL, 1964). The dispersion of totals from the mean value of annual rainfall is shown in Fig.8. The standard deviation ranges from more than 400 mm over Kwangsi and the southernmost portions of Kwangtung and Taiwan to only 200 mm north of the Yangtze. The standard deviation of annual rain-days, which is influenced much more by local atmospheric instability than by the frequency of typhoons, is 26 days in Hunan and decreases to 14 days on the Kwangtung coast, over Taiwan, Liaoning, Kansu and the Red Basin.

Seasonal variation of rainfall

Except in Sinkiang and northeast Taiwan, the seasonal variation of rainfall is characterised by a pronounced summer monsoon maximum and a comparatively dry winter. In most places and more particularly in the south, the wettest months are approached by a gradual increase in rainfall during March, April and May, corresponding to the increase of instability and moisture content in the atmosphere. In the mountainous country from Tasueh Shan westwards, 75–95% of the annual total is recorded during May to September (Lu, 1947), and throughout Hopei 90% of the annual total occurs within this period. Over the rest of the country the seasonal variation is less marked. Along the Yangtze and over the southeastern provinces, the maximum occurs most often in June and is associated with small depressions passing seawards along a trough lying east–west somewhere between the Yangtze and the southern coast. These are the "Mai-yü" depressions and the early rains of April, May and June are often called the "Mai-yü" or "Plum" rains. Elsewhere, maxima occur at various times during the summer, but most frequently in July or August (Fig.9) when the contribution of tropical cyclones is more pronounced.

Each year the heavy monsoon rains do considerable damage. Lives are lost in landslides, the fertile soil is stripped from the slopes and crops on the lowlands are buried by flooded rivers. MIN TIEH (1941) estimates that the principal rivers of the plains discharge 665 million m³ of valuable silt into the ocean every year. Over many years, the Hwang Ho has gradually built up its bed by the deposition of sediment and it now flows between man-made dykes above the general level of the plain. On many occasions after continual heavy rains, the river has broken through the dykes to destroy cultivation over vast areas and

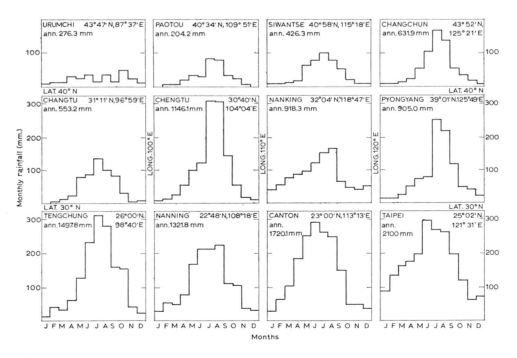

Fig.9. Mean monthly rainfall at selected stations in Central and East Asia.

take millions of lives. The course of the river is constantly changing and it now enters the sea 400 km further north than it did a century ago. Flooding in the Yangtze is less frequent but it also can be severe. Every summer the river rises about 12 m at Wuhan; but following the passage of several tropical cyclones in July–August 1931, the level rose 16 m and seriously flooded an area of 9 million hectares.

Rainfall decreases rapidly through late September and October as successive outbursts of cold dry air pass southward across the country.

In Sinkiang, seasonal variations are less marked. Urumchi exhibits a slight maximum in October, while in other parts of Dzungaria cold fronts penetrating the mountain-passes create an indistinct maximum in winter (LU, 1944). However, to the south of Tien Shan in the Tarim Basin, C. Y. CHANG's (1949) observations indicate a spring instability maximum with 60–70% of the annual falls at Kucha, Khotan and Kashgar occurring between March and May. In northeast Taiwan, which is sheltered from the summer souther-lies and exposed to the northeasterlies, most rain falls in winter.

KENDREW (1941) distinguishes three major rainfall regimes in China—those of the Yang-tze, northern China and southern China. Peking is typical of the north with about 600 mm/year; 90% of the total amount occurs between May and September while each of the remaining months has less than 20 mm. The south has a similar regime but heavier annual rainfall, heavy summer rains, an early summer maximum and unimportant winter rains. Hong Kong is typical of the south with over 2,000 mm/year, 25 mm or more each month, more than 250 mm each summer month and a maximum in June. In the central regime most rain falls in the summer monsoon with a maximum in June or July, but the distinction between the seasons is less marked than in the other regimes and winter rainfall is quite considerable. Thus, at Shanghai each month has more than

TABLE XII

LU'S RAINFALL REGIONS

Type	Description	Seasonal distribution of rainfall (%)			
		winter	spring	summer	autumn
East Hainan	September maximum, July–August wet	7	20	42	31
Tongking Gulf	Rapid increase to July–August, abrupt decrease	5	19	56	20
Nanling	Two maxima, May and August	10	32	45	13
Southeast coast	Two maxima, June and August–September	10	24	48	18
Northeast Taiwan coast	Maximum November–March; peak December	31	26	16	27
Yangtze delta	Two maxima, June–July, and April	14	24	44	18
Middle Yangtze	Similar	12	36	37	14
North China (north of Chinling)	Rapid increase July–August; abrupt decrease	3	10	70	17
Southwest China	Two maxima, late spring and early autumn	5	26	41	28
Tarim	Early summer rain			40% in June	
Dzungarian	Winter maximum	37		15	

30 mm and only 60% of the total amount occurs between May and September. Lu (1944) distinguishes eleven different seasonal rainfall regimes over China as shown in Table XII.

All Lu's regions exhibit a summer monsoon maximum, except those of Dzungaria where winter cold fronts are important and of northeastern Taiwan where orography plays a prominent part. The winter maximum of Taiwan is confined to the narrow coastal strip most exposed to the monsoon northeasterlies which have picked up moisture over the Yellow and Eastern Seas; summer rainfall on this coast is small because of sheltering from the moist monsoon southwesterlies.

Lu considers that the most important factor contributing to the September maximum over the eastern Hainan coast is typhoon rainfall, to which the coast is more exposed than places within the Gulf of Tongking. In support of this, CHIN (1958) has shown that, during a period of 70 years, 12 typhoons crossed eastern Hainan in June, 27 in July, 32 in August, 47 in September and 17 in October. Lu's other climatic divisions are distinguished by the occurrence of double rainfall maxima at different times in different regions. Thus, the two Yangtze climates are said to have maxima in June and July and again in April. However, although April rainfall is not inconsiderable in some places along the Yangtze, an April maximum is not readily discernible in the statistics available now. Long-period data for places on the Yangtze indicate maxima in June at Shanghai, August at Nanking, May and September at Wuhu, June and August at Anking, June at Wuhan, July and May at Kiangling, September and June at Wanhsien, and June and October at Fowling.[1] Similar difficulties arise when an attempt is made to define other climatic regimes according to the time of occurrence of double maxima.

Diurnal variation of rainfall

Near the coast, the summer rainstorms are sometimes very intense. At Hong Kong, falls of 101 mm and 256 mm have been recorded in periods of 1 and 3 h respectively, while at Paishih[1] falls of 437 mm, 771 mm and 1,248 mm have been recorded in 6, 12 and 24 h (PAULHUS, 1965).

The time of day at which rain most frequently occurs varies with locality and season (WATTS, 1962). Over most of the region and for the greater part of the year, the maximum frequency is in the afternoon. However, during May and June the maximum occurs in the morning along the Chinese coast and as far north as Korea (RAMAGE, 1952). This effect does not extend more than 60 miles inland, and does not reach Lungchuan, Lungyen or Hoyun.[1] It is found at Tainan but not at Taipei where there is always a maximum in the afternoon. By August, morning maxima are replaced generally by afternoon maxima, except very close to the coast near Hong Kong and Macau where morning maxima persist throughout the summer. During May and June, a night or early morning maximum frequency of precipitation is also found in the high country west of Szechuan, Chinghai and Yunnan (HANSON-LOWE, 1941; LUI, 1957; WATTS, 1962).

---

[1] Wuhu: 31°23′N 118°25′E, Anking: 30°34′N 117°03′E, Kiangling: 30°20′N 112°15′E, Wanhsien: 30°54N 108°20′E, Fowling: 29°44′N 107°22′E, Paishih: 24°33′N 121°13′E, Lungchuan: 24°03′N 115°14′E, Lungyen: 25°10′N 117°00′E, Hoyun: 23°41′N 114°45′E.

Snow

Precipitation of any sort is generally sparse in the north, but in the northeast and as far south as Shanghai the colder spells are ushered in by violent snow storms. Mean charts published by the Central Weather Bureau (ANONYMOUS, 1955) show that snow first appears in Heilungkiang and Kirin at the beginning of October and occurs on an average of 30–40 days up to the last fortnight in April. Southwards, there is a decrease both in the frequency of snowfalls and in the length of the season during which they occur. Over Liaoning and North Korea there are 20 days of snowfall in the period November to early April, but there are less in South Korea where the season extends only from early December to mid-March (ANONYMOUS, 1962a). Over central China from Peking to Hunan and Kiangsi, snow falls on 10 days between the beginning of December and mid-March, but the frequency decreases sharply towards the south and southeast. There are less than 5 days per year during January or February on the Chekiang coast south of Shanghai, while along 25°N latitude there is on the average only one fall per year occurring in the last week of January. In Taiwan there are two or three January or February falls each year in the mountains, but these do not extend to low levels.

In the low-lying land surrounding the Red Basin there are normally only 5 days of snow per year and they are confined to January; but throughout the higher country of western Szechuan, snow falls on 20 days between the last week of October and the beginning of May. Notes by HANSON-LOWE (1941) indicate a more complex regime west of Tasueh Shan while Central Weather Bureau mean charts (ANONYMOUS, 1955) show Changtu Town to have 18 snow-days between mid-October and mid-April, and Lhasa (Tibet) to have only 8 days between mid-October and the end of April.

Around the edges of the Tarim Basin and in the Turfan, there are minor falls of snow on 4–8 days between mid-November and early March each year, but in Dzungaria just to the north of Tien Shan, falls are recorded on 30–40 days between mid-October and mid-April. Snow is more frequent in the mountains of Sinkiang and this plays an important part in the economy of the arid Tarim; C. Y. CHANG (1949) estimates that 80% of the water used for irrigation in Kashgar comes from mountain snow and 20% from springs.

Snow cover

The average duration of snow cover in different places is shown in Fig.10. The ground is more than half covered with snow for an average of 150 days throughout October to April in the northernmost parts of Heilungkiang and Inner Mongolia, and for about 100 days or more during November–March in Kirin. In Korea there is a large variation of 100 days during November–March in the north to less than 5 days during December–February on the southeastern coast. Over central China the duration of cover amounts to only 10 days, between mid-December and the beginning of March in Hopei and between mid-January and mid-February in Hupeh and Anhwei. The duration of snow cover decreases southwards to one occasion annually about the beginning of February along 25° latitude but also to only one occasion annually in the last fortnight of January in the Red Basin. On the high country of western Szechuan, snow cover persists for 25 days or more between mid-November and mid-March. Further to the west there is considerable variation with altitude and exposure; Changtu Town at 3,200 m has only 10 days

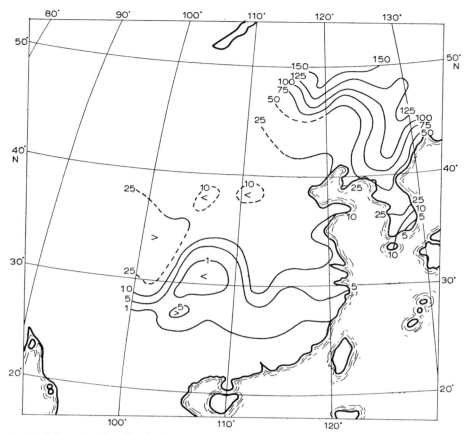

Fig.10. Mean duration (days) of snow cover each year over Central and East Asia.

of snow cover each year between about mid-November and the latter half of March. Around the Tarim Basin and the Turfan Basin, the snow cover persists for only short periods ranging from 4 to 18 days starting about the end of November and finishing early in February. However, just to the north of Tien Shan, the cover apparently lasts much longer, as Kuldja (Ining) averages 110 days between early November and mid-March and Urumchi has 144 days between mid-October and mid-April.

Hail

According to GHERZI (1951) hail is not very frequent in China except in Kansu and Shensi. More recent data presented by the Central Weather Bureau (ANONYMOUS, 1955) indicate that, although hail occurs once or twice a year in Shantung, Anhwei, Kweichow and Yunnan, it is otherwise practically unknown on the plain, in the southeastern provinces, Taiwan or the arid northwest. At altitudes above 1,000 m to the north of the North China Plain and west of Tung Ting Hu there are at least 1–2 hailstorms a year, with a maximum of about 4 on the Ordos Plateau and in Kirin, and at least 6 in the Tasueh and Bayan Kara Shan. On the plateau west of Tasueh Shan the frequency of hail appears to vary considerably with locality, and HANSON-LOWE (1941) draws attention to conditions at Kantse and Taofu which are at about 3,000 m M.S.L. and 140 km apart. Hail is rare at Kantse, but at Taofu there are frequent intense hailstorms during the wet season from May to October.

Thunderstorms

Mean maps of thunderstorm days (or days on which thunder is heard) published by the World Meteorological Organization (Anonymous, 1956) show a simple pattern with greatest frequency of more than 40 per annum over Changtu Province and over Hainan, decreasing northeastward to less than 10 days over Liaoning, Kirin and Heilungkiang. Within these limits, there are generally about 30 days annually over the southern and western provinces and 10–20 days from Fukien and Hupeh to Hopei. These frequencies are small compared with the average of 140 days recorded in Malaya.

Mean maps published by the Central Weather Bureau (Anonymous, 1955) give more detail and show rather higher frequencies than the World Meteorological Organization publication (Anonymous, 1956). Hainan and the Luichow Peninsula are shown to have 90–100 thunderstorm-days per annum all of which occur during March to September. Otherwise 50 thunderstorm days are recorded between mid-February and the end of September each year to the south of the Yangtze, during March-September in the central ranges of Taiwan and from mid-March to early October in the high country south and west of the Tanglha and Ning Ting Shan. From the Yangtze northward and east of 100°E longitude there are about 20 days each year, the season lasting from mid-April to late September along the Yangtze and from May to late September in the far north. In Sinkiang the season lasts only from mid-May to the end of August, and frequencies range from 5 days annually in the deserts to 20 in the mountains.

Temperature, humidity and evaporation

Mean annual temperatures

Mean annual temperatures vary from −4°C in northernmost Heilungkiang and 5°C in Sinkiang and West Szechuan to 23°C on the southern Chinese coast and in Taiwan. The annual 20° isotherm lies along 25°N latitude and there is a general decrease of temperature northward on the plain. The 10° isotherm, which is mainly along the eastern edge of the Tibetan Plateau below the 1,000 m contour, lies to the west of the Red Basin and close to Lanchow, Taiyuan and Peking. In the extreme northeast, the 0° annual isotherm is along 48°N latitude. In Korea, annual temperatures range from less than 5°C inland at 42°N latitude to more than 12°C on the southern and southeastern coasts (Anonymous, 1962a).

The annual mean masks a large seasonal range in the monthly means amounting to 50°C in the north and northwest, 25°C on the plain and 15°C on the south China coast. Lu (1944) considers that seven distinct types of seasonal temperature ranges can be distinguished in different parts of China.

Winter temperatures

Throughout the winter, the cold monsoon northerlies or northeasterlies are very persistent and January is the coldest month in practically the whole region. Under the influence of successive colder and more vigorous outbursts from the Mongolian anticyclone,

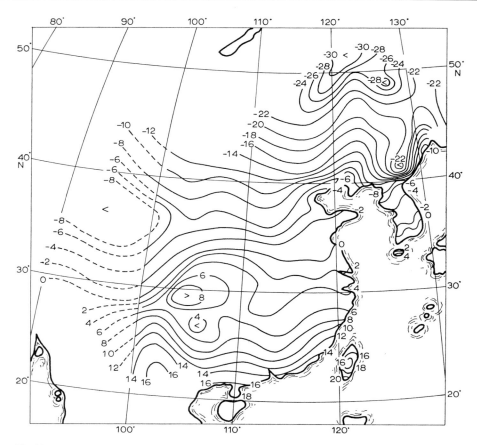

Fig.11. Mean air temperature (°C) over Central and East Asia during January.

temperatures in the north decrease rapidly during October and this effect extends to the south after mid-November. By January the isotherms lie latitudinally with mean temperatures of −30°C north of latitude 50°N, −20°C on latitude 45°N and −10°C on 40°N (Fig.11)[1]. In this region the 0°C isotherm reaches nearer to the equator than anywhere else in the world, lying mainly along 33°N latitude on the plain but curving around to the south of the Himalayas. January temperatures are 10°C along 25°N latitude and are as high as 20°C on the southern parts of Taiwan and Hainan.

North of Po Hai and of 40°N latitude, mean monthly temperatures are generally below −10°C throughout December to February, while mean maxima are below 0°C, mean minima below −15°C and extreme minima below −40°C. Daily mean temperatures of 0°C or less are recorded on 100–150 days each year and persist till the beginning of March (Fig.12). The land is snow covered and rivers are frozen for six months of the year, while broken ice extends out far beyond Po Hai.

In Sinkiang and the mountainous west, winter temperatures average −10°C with a diurnal range from −5°C to −20°C and with at least 200 days when temperatures are 0°C or

---

[1] Fig.11 and 13 are based on uncorrected air-temperatures at a large number of observing stations. In general, the isotherms represent temperatures at the average elevation of that part of the country, but the scale of the map does not permit showing variations of temperature with height in the mountainous regions.

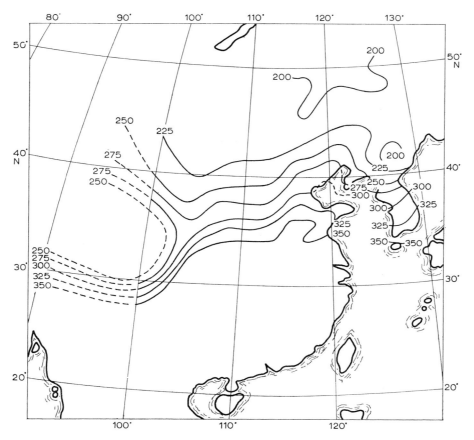

Fig.12. Mean number of days each year with daily average temperature greater than 0°C over Central and East Asia.

lower. On the North China Plain the winters are less rigorous, with mean montly temperatures of close to 0°C in December–February, and with monthly maxima and minima of about 4° and −4°C respectively. Ground frosts are frequent but daily mean air temperatures fall to 0°C on less than 50 occasions annually and rarely after the end of January (Fig.12).

Just to the south of the Yangtze, mean temperatures are about 3°–5°C in January and February and there are a few frosts. Conditions are rather better in the Red Basin, however, where sheltering maintains average temperatures of 6°–10°C throughout the winter. Further to the south, winter temperatures are above 14°C with daily maxima approaching 20°C, but on occasion temperatures down to freezing have been reported as far south as Hong Kong (PEACOCK, 1952).

Summer temperatures

During March the cold northerly outbursts become weaker and less frequent, until by the middle of April very few of them reach the southern provinces. Instead, increasingly frequent incursions of moist tropical maritime easterlies tend to raise temperatures throughout the entire region. Mean daily temperatures rise above 6°C on the North China Plain by mid-March, in western Szechuan and Sinkiang by the beginning of April, at Har-

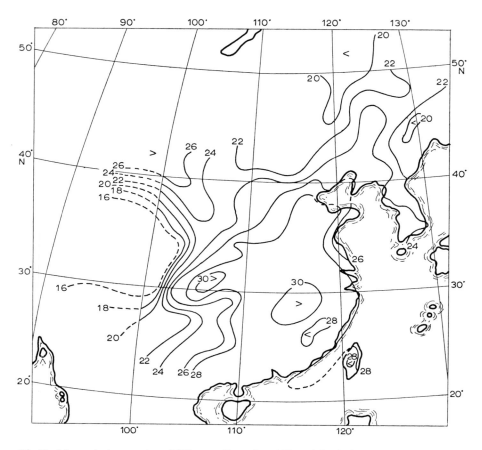

Fig.13. Mean air temperature (0°C) over Central and East Asia during July.

bin by mid-April and over the northernmost parts of the region by May (ANONYMOUS, 1953). By May, mean temperatures are generally above 20°C at low levels south of the Hwang Ho and above 10°C further to the north. Although there is no regular advance of a monsoon front as in India, temperatures rise very rapidly over the whole country from March to May. In January, temperatures in north China are 50°C lower than those in the south, but by June the difference is less than 15°C and temperatures are generally above 16°C.

August is the hottest month in Korea and in a few other places in the east, but otherwise mean monthly maximum temperatures are greatest in July. Daily temperatures are 18°C or higher for more than 50 days each year in the extreme north, 150 days on the North Plain, more than 200 days in sheltered Szechuan and to the south of 25°N latitude and 350 days in the southern parts of Taiwan and Hainan (ANONYMOUS, 1953). Daily temperatures are as high as 25°C on at least 50 days each year south of Peking, on more than 100 days south of 25°N latitude and for 175 days in southern Taiwan and Hainan. Highest temperatures are observed in valleys well inland, and P. K. CHANG (1934) considers that Turfan and the mid-Yangtze are the hottest parts of China. The absolute maximum temperatures recorded in China are generally over 38°C and exceed 45°C in some places far from the coast, but in the southwest and south they are between 30° and 35°C.

Summer monsoon conditions continue throughout September, but cold outbursts start to affect the far north in October. The net effect of successive cold waves is that from

September up till December when winter conditions are fully established, mean temperatures fall 10°C each month in the northern provinces and 5°C each month in the south.

Departures from the mean temperatures

Although the seasonal range of temperature is large, conditions do not vary greatly from one year to another. The greatest departures of annual totals from the normal rarely exceed 1°C in the south or 2°C in the north. Summer temperatures are equally uniform and the standard deviation of July totals exceeds 1.5°C on the central plain only (Fig.14). Winter temperatures show a slightly larger deviation but monthly means are still mainly within 2°C of the normal except to the north of 40°N latitude (Fig.15).

Relative humidity

Prolonged subsidence in the Mongolian anticyclone maintains a very dry atmosphere throughout the winter. At Irkutsk, which is not far from the usual location of the centre of the anticyclone, mean winter temperatures and dew-points are below −20°C and mean relative humidity is 80% or less (Table XIII).
Further south at Paotou on the Ordos Plateau, relative humidity is a little over 60% during the winter and dew-point temperature is below −15°C. As the cold air flows down

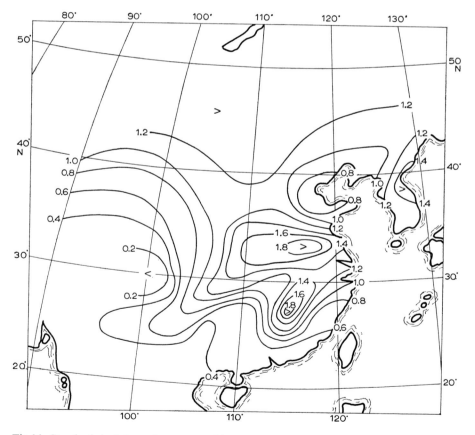

Fig.14. Standard deviation of mean monthly temperatures (°C) in July.

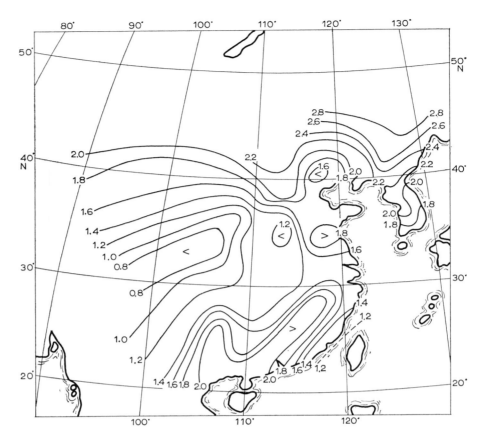

Fig.15. Standard deviation of mean monthly temperatures (°C) in January.

southwards on to the North China Plain it is heated adiabatically and by contact with the warmer ground so that relative humidity is decreased considerably. At Peking the average humidity is as low as 50% throughout the winter. It is temporarily increased by rain as each cold front passes southwards, but falls to very low values when a steady monsoon northerly is blowing.

South of Shantung and of 35° latitude mean humidities are higher and are generally over 75% in winter (Table XIII), because the cold air often approaches from a northeasterly direction and thus gains some moisture in the lower layers from the sea. However, there are many occasions when vigorous northerly outbursts reach south China without deviation and at such times extremely low humidities are recorded. For example, early in January 1959 the Mongolian anticyclone intensified very rapidly until central pressures reached over 1,050 mbar. The result was that a direct north–northwesterly outflow quickly swept from Inner Mongolia to the China coast where dew-point temperatures fell to below −10°C and humidity to less than 10% (*Meteorological Results, Surface Observations, 1960*, Royal Observatory, Hong Kong).

During March and April, increasingly frequent incursions of maritime air bring about a small increase of humidity in coastal areas; but further inland the humidity decreases slightly with the seasonal rise of temperature.

The summer is humid and enervating over practically the whole country and particularly in river-valleys such as the Yangtze. North of the Yangtze, dew-point temperatures

TABLE XIII

TEMPERATURE, RELATIVE HUMIDITY AND DEW-POINT TEMPERATURE AT SELECTED STATIONS

	Month	J	F	M	A	M	J	J	A	S	O	N	D	Year
Irkutsk	Mean temperature (°C)	−21	−18	− 9	2	9	15	18	15	8	1	−11	−19	− 1
	Relative humidity (%)	80	74	68	59	56	66	74	78	78	75	80	85	73
	Dew-point temperature (°C)	−23	−21	−14	− 6	0	9	13	11	5	− 3	−14	−20	−5
Paotou	Mean temperature (°C)	−12	− 9	0	7	15	21	23	21	15	8	− 2	−10	6
	Relative humidity (%)	63	64	49	48	49	52	64	69	61	59	61	61	58
	Dew-point temperature (°C)	−17	−15	− 9	− 3	5	11	15	15	7	0	− 8	−16	− 1
Peking	Mean temperature (°C)	− 5	− 2	5	14	20	25	26	25	20	13	4	−3	12
	Relative humidity (%)	50	50	48	46	48	56	72	71	67	59	56	51	56
	Dew-point temperature (°C)	−13	−11	− 6	2	9	15	21	19	13	5	−4	−12	3
Wuhan	Mean temperature (°C)	4	5	10	16	22	26	29	29	24	18	12	0	17
	Relative humidity (%)	76	78	76	77	75	77	77	75	74	73	75	74	76
	Dew-point temperature (°C)	0	2	6	12	17	21	25	24	19	13	8	− 4	13
Hong Kong	Mean temperature (°C)	15	15	17	21	25	27	28	28	27	25	21	17	22
	Relative humidity (%)	75	79	83	85	85	84	83	84	79	72	69	70	79
	Dew-point temperature (°C)	11	12	15	19	22	24	25	25	23	19	15	12	19

TABLE XIV

MONTHLY AND ANNUAL TOTAL EVAPORATION (MM) AT SELECTED STATIONS DURING 1960

Place	J	F	M	A	M	J	J	A	S	O	N	D	Year
Harbin	9	51	93	219	215	187	217	149	158	111	47	10	1466
Urumchi	11	15	29	166	218	251	269	309	201	128	34	10	1641
Peking	39	98	122	257	291	284	195	160	163	104	70	56	1839
Shanghai	47	78	72	124	148	146	262	188	153	147	70	60	1495
Changtu	57	81	147	205	236	162	168	185	116	106	78	57	1598
Wuhan	45	74	75	100	139	180	241	272	156	136	70	63	1551
Chengtu	44	50	79	93	142	116	142	115	91	64	47	35	1018
Foochow	88	112	92	109	105	151	256	170	166	204	112	101	1666
Hong Kong mean evap. 5 years	116	90	111	122	155	160	182	170	168	187	148	128	1737
Hong Kong pot. evap. 5 years	79	76	86	97	128	127	146	150	138	137	113	89	1366

are about 20°C in July and August and average daily relative humidity exceeds 70% except in Sinkiang where the air is very dry. South of the Yangtze, dew-points are generally about 25°C and humidity is greater than 80%. Humid conditions are maintained until the arrival of the first bursts of cold northerly air in October and November.

Evaporation

Table XIV contains data from a few of the 188 stations in China where evaporation observations were made during 1960. It shows that evaporation is greatest in summer and that the amplitude of its seasonal variation decreases southwards and towards the coast. An outstanding feature is that, despite the low winter temperatures in the north, mean annual evaporation there is practically as great as on the tropical south coast. This is no doubt related to the clear skies and the great duration of sunshine (Fig.16) in the northern–central provinces and deserts where summer monsoon clouds and rains infrequently penetrate. In fact the total duration of sunshine at Paotou during June to September is 30% higher than the total for Canton.

J. H. CHANG (1955) has computed annual potential evapotranspiration from the climatological data of 285 stations covering all China except Sinkiang, Chinghai, Changtu and western Szechuan (Fig.17). The isopleths of evapotranspiration form a pattern broadly

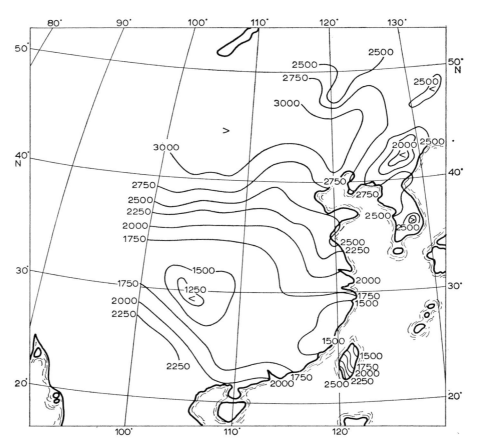

Fig.16. Mean duration (h) of sunshine each year over Central and East Asia.

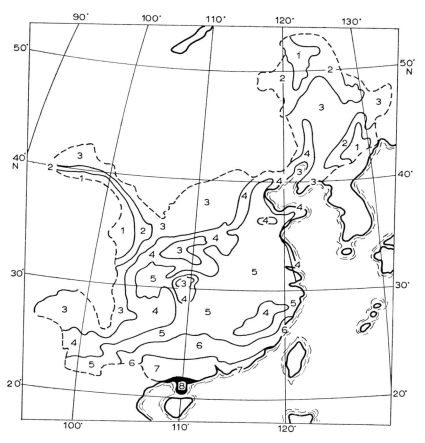

Fig.17. Annual potential evapotranspiration regions over China, excluding Sinkiang, Chinghai and Changtu. (After J. H. CHANG, 1955.)

The potential evaporation of the various regions is as follows: region 1: 285–427 mm; region 2: 427–570 mm; region 3: 570–712 mm; region 4: 712–855; region 5: 855–997; region 6: 997–1,140 mm; region 7: 1,140–1,282 mm; region 8: 1,282–1,425 mm.

resembling that of the mean annual isotherms, with a maximum on the south China coast and minima in the northwest and north and on the North Korean border. Mean annual potential evapotranspiration actually measured at Hong Kong (*Meteorological Results, Surface Observations, 1959*, Royal Observatory, Hong Kong) is 1,366 mm which is a little higher than Chang's estimate.

Chang shows that over the greater part of China there is a large average water surplus which exceeds 800 mm in Fukien and Chekiang. North of the Hwang Ho and west of 120°E longitude there is a deficiency mainly over 200 mm but exceeding 600 mm in Kansu. These features can be seen from the regional distribution of moisture index (Fig.18). Moisture index is taken to be $(100s-60d)/n$ where s is the water surplus in the rainy season, d is the deficiency in the dry season and n is the water need (THORNTH-WAITE, 1948). The line separating regions 1, 2 and 3 from regions 4–9 in Fig.18 marks the boundary between the dry north and the moist south and approximately corresponds with THORP'S (1936) boundary between the pedocal and pedalfer soils. The only perhumid climate on the map is found in the Nanling Hills.

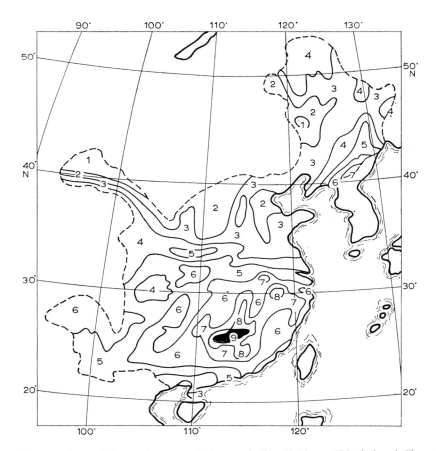

Fig.18. Moisture index regions over China, excluding Sinkiang, Chinghai and Changtu. (After J. H. CHANG, 1955.)

Region	Moisture index	
1	< -40	arid
2	-40 to -20	semi-arid
3	-20 to 0	dry sub-humid
4	0 to 20	moist sub-humid
5	20 to 40	
6	40 to 60	humid
7	60 to 80	
8	80 to 100	
9	>100	perhumid

Atmospheric obscurity

Fog inland

The frequency of occurrence of fog[1] varies considerably from one part of China to another. It is rare in the west or the arid north, but occurs very frequently in the sheltered

---

[1] A day of fog is defined as one in which horizontal visibility falls below 1,000 m.

TABLE XV

AVERAGE NUMBER OF DAYS OF FOG IN EACH 3-MONTH PERIOD

Place	Dec. Feb.	Mar. May	Jun. Aug.	Sep. Nov.	Year	Period
Urumchi	21	4	0	6	31	1908–1930, 1951–1953
Harbin	5	2	4	5	16	1909–1936, 1949–1952
Tientsin	12	3	3	8	26	1946–1952
Paoting	13	5	3	20	41	1949–1952
Taiyuan	7	1	2	7	17	1926–1937
Sian	18	6	6	21	51	1951–1953
Peking	7	3	6	5	21	1925–1936,
Chengtu	38	22	23	28	111	1932–1952
Chungking	39	33	30	32	134	1941–1953
Kweiyang	23	20	25	27	95	1936–1953
Nanking	14	8	3	9	34	1929–1936, 1951–1953
Wuhan	21	11	0	12	44	1950–1953
Changsha	15	13	5	12	45	1933–1937, 1947–1955
Nanchang	12	6	2	7	27	1936–1937, 1951–1953
Kweilin	23	14	7	12	56	1938–1942, 1950
Nanning	3	5	1	3	12	1951–1955
Dairen	3	10	14	1	28	1905–1940, 1950–1952
Tsingtao	7	16	23	2	48	1902–1952
Howki Island	13	18	28	5	64	15 years
NE Shantung Promontory	18	27	53	5	103	15 years
Shanghai	9	10	6	11	36	1883–1911, 1951–1953
Hangchow	14	11	6	18	49	1934–1937, 1951–1952
Gutzlaff Island	21	35	23	10	89	1926–1934, 1950–1955
Steep Island	20	41	35	10	106	15 years
Peivushan	18	39	26	9	92	15 years
Wenchow	7	9	2	2	20	15 years
Amoy	3	8	1	0	12	1951–1953
Canton	20	18	6	10	54	1925–1936, 1950–1952
Hong Kong	5	10	0	1	16	1953–1955
Foochow	12	19	9	6	46	1951–1955
Taipei	5	4	2	2	13	1897–1937
Tainan	6	3	0	3	12	1897–1937
Pehta Tao	16	33	12	9	70	15 years
Nanpeng Tao	12	28	17	5	62	15 years
Breaker Point	11	25	12	5	53	15 years
Waglan Island	11	24	1	0	36	1919–1933

TABLE XVI

PERCENTAGE OF REPORTS OF FOG AT EACH OBSERVATION HOUR (TOTAL PERIOD OF OBSERVATIONS 1957–1959 FOR ALL PLACES EXCEPT HONG KONG AND WAGLAN 1955–1962)

Standard time	Dec. Feb.	Mar. May	June Aug.	Sep. Nov.	Dec. Feb.	Mar. May	June Aug.	Sep. Nov.	Dec. Feb.	Mar. May	June Aug.	Sep. Nov.
	Tsanghsien				*Nanking*				*Matsu*			
02	14	1	2	4	17	12	6	14	4	14	4	0
08	20	5	12	3	34	21	14	28	4	15	5	1
14	8	0	0	0	3	1	1	1	2	9	1	0
20	11	1	0	1	8	3	1	5	5	9	3	0
	Tzeyang				*Wuhan*				*Quemoy*			
02	8	0	0	0	18	9	3	8	3	13	1	1
08	14	13	6	6	32	30	10	22	6	24	2	2
14	7	2	0	0	7	2	0	0	2	5	0	0
20	7	1	0	0	11	2	0	5	3	9	0	0
	Loyang				*Lotien*				*Waglan*			
02	6	4	1	3	2	1	1	2	3	9	0	0
08	29	22	9	24	8	2	13	20	3	7	0	0
14	12	5	0	1	0	0	0	0	3	4	0	0
20	4	1	0	1	0	0	0	0	3	4	0	0
	Luchow				*Canton*				*Hong Kong*			
02	29	6	1	11	2	3	0	2	1	1	0	0
08	43	18	13	29	12	13	4	6	3	1	0	0
14	33	2	0	6	0	0	0	0	0	0	0	0
20	26	1	0	5	0	1	1	1	0	0	0	0

valleys of Szechuan and Kweichow and just offshore in the China and Eastern Seas. Over most of Sinkiang there is little fog, and at Hami, Turfan and Kucha, it only occurs on about four days annually. However, it forms more frequently to the north of Tien Shan, and Urumchi has about 31 fog days each year most of which are between November and March (Table XV).

Although cloud frequently envelops the higher peaks of the mountainous southwest provinces, fog is hardly ever reported at Changtu Town or Kantse. Similarly in the northern highlands, Paotou and Hsining and Lanchow rarely have more than four days of fog in a year. Frequencies are a little higher inland in the northeastern provinces; Harbin and Changchun have 16 and 6 days respectively without any pronounced seasonal variation.

Over the land from Hopei southwards, fog occurs on 20–50 days per year. On the North China Plain, about 70% of these occasions are in the period September–January (see Tientsin, Paoting, Taiyuan and Sian in Table XV) while April, May and June are comparatively free. The fogs of northern and central China are radiation fogs with a maximum about sunrise as shown in Table XVI for Tsanghsien, Tzeyang and Loyang[1].

[1] Tsanghsien: 38°19'N 116°54'E, Tzeyang: 35°35'N 116°53'E, Loyang: 34°47'N 112°26'E.

During September to December the fog normally dissipates during the morning, but in January it is very persistent and often lasts till after midday.

Highest frequencies are recorded in the enclosed valleys of the interior, and particularly in the Red Basin where fog occurs on one out of three days without any marked seasonal variation (see Chengtu, Chungking and Kweiyang in Table XV). In the valleys of the interior it often persists throughout the day as shown in Table XVI for Luchow during December–February.

From the Yangtze southwards the seasonal variation becomes less pronounced, though in general fog occurs least during June to August (e.g., Nanking, Wuhan, Changsha, Nanchang, Kweilin and Nanning in Table XV). Towards the south the diurnal variation becomes more pronounced and, except in the deep river valleys, the fog invariably dissipates during the day (see Nanking, Wuhan and Lotien[1] in Table XVI).

Fog on the coast

On the coasts bordering Po Hai and the Yellow Sea there is a very different regime, with fog occurring mainly in spring and summer but rarely in autumn. Thus, at Dairen and Tsingtao 75% of the fog-days occur between March and July. Statistics from GHERZI (1951) for Howki Island and Shantung Northeast Promontory[1] show rather higher annual totals on this part of the coast but they exhibit the same seasonal characteristics. At many places near the coast in Kiangsu and North Chekiang, fog occurs throughout the year but less in summer than in other months (e.g., Shanghai and Hangchow in Table XV). However, offshore at Gutzlaff Island, Steep Island and Peivushan[1] where annual totals are higher, there is a very distinct seasonal maximum with about 75% of the fog-days occurring during March to July.

Along the coast from Wenchow southwards, fog-days range from about 10 to 50 a year and there is a very distinct seasonal variation with maxima during January–April and minima during the summer. (See Wenchow, Amoy, Hong Kong, Taipei and Tainan in Table XV). Just offshore there are about 40–70 days per annum as shown by the statistics for Pehta Tao, Nanpeng Tao, Breaker Point and Waglan Island[1]. At places a few miles in from the coast (e.g., Canton, Table XVI) the fog clears each morning and appears to be of local origin, but observations from Matsu, Quemoy and Waglan Island indicate that fog over the sea is much more persistent[1]. This sea-fog or "crachin" develops in a moist tropical stream of air passing from the warm Kuroshio Current to the much colder waters near the China coast (RAMAGE, 1954). HUNG (1951) has shown that during March and April the fog at Waglan Island clears during the afternoon on only 50% of the total number of occasions. Hung has also shown that the edge of the fog-bank moves into Hong Kong harbour when the wind is favourable, though Table XVI indicates that such incursions are mainly confined to the mornings.

Dust and sand in the atmosphere

One of the most notable characteristics of late winter and spring in Southeast Asia is the

---

[1] Lotien: 25°25′N 106°49′E, Howki Island: 38°04′N 120°38′E, Shantung North Promontory: 37°24′N 122°42′E, Gutzlaff Island: 30°49′N 122°10′E, Steep Island: 30°13′N 122°35′E, Peivushan: 28°52′N 122°14′E, Pehta Tao: 25°58′N 119°59′E, Nanpeng Tao: 23°15′N 117°18′E, Breaker Point: 22°56′N 116°22′E, Waglan Island: 22°01′N 114°18′E, Matsu: 26°10′N 119°59′E, Quemoy: 24°26′N 118°20′E.

41

TABLE XVII

NUMBER OF DAYS OF SAND-STORMS (COMPILED FROM 3-YEARS' DATA DURING EITHER 1951–1953 OR 1953–1955).

Place	Dec. Feb.	Mar. May	Jun. Aug.	Sep. Nov.	Year	Place	Dec. Feb.	Mar. May	Jun. Aug.	Sep. Nov.	Year
Paotou	3	11	3	1	18	Peking	0	3	0	0	3
Paoting	3	8	6	1	18	Changchun	1	2	0	0	3
Taiyuan	4	9	2	0	15	Tientsin	0	2	0	0	2
Lanchow	0	7	4	0	11	Changtu	6	4	0	3	13
Sian	0	3	1	0	4	Kantse	18	13	1	3	35
Changsha	0	0	1	0	1	Turfan	1	9	2	1	13
Tsinan	0	4	1	0	5	Hami	0	17	10	6	33
Tsingtao	0	2	0	0	2	Urumchi	1	2	0	0	3

large amount of dust and sand in the atmosphere. Through the centuries the monsoon northerlies have been sweeping up solid matter from the rainless northern deserts and carrying it across the North China Plain, and this process has probably contributed much to the vast loess deposits at Shensi and Shansi. The loess, which reaches a depth of 100 m in places, extends southward to the Chinling Hills and even further over the plain through Honan. In our time, most of the material lifted from the Gobi Desert during winter is probably coarse sand, but fine dust is still being regularly carried from the Ordos to the plain and thence to the southernmost parts of the country.

Over northern Kansu and that part of Inner Mongolia which lies to the west of 110°E longitude, sandstorms occur on about 30 days each year[1] (ANONYMOUS, 1955). Some of the worst sandstorms are associated with cold fronts and the frequency of such occasions decreases to the south and east. Nevertheless, there are at least 10 such days annually

TABLE XVIII

THE PERCENTAGES OF TIME DURING EACH MONTH WHEN VISIBILITY AT HONG KONG EXCEEDS 17 KM[1]

Month	% of time
January	31
February	21
March	22
April	19
May	34
June	45
July	63
August	52
September	49
October	55
November	47
December	31

[1] Computed from hourly observations over 5 years 1958–1962.

[1] A sandstorm day is considered to be one during which the visibility is limited by flying sand or dust to 1000 m or less.

along the mountainous northern boundary of Chinghai and in the provinces of Shansi, Honan and western Hopei. More specifically, there are 18 days of sandstorms annually at both Paotou and Paoting, 15 at Taiyuan and 11 at Lanchow (Table XVII).

Although these statistics represent extreme cases, some dust or fine sand is nearly always present in the northern atmosphere during late winter and spring and is found at levels up to 4,000 m above the ground (GHERZI, 1951). In the north, dust enters dwellings despite closed windows and restricts visibility to a few km except immediately following rain.

Further to the south and southeast, sandstorms become fewer, with four per year, a Sian, one at Changsha, five at Tsinan and two at Tsingtao. There is a similar decrease eastward, with only three days per year at Peking and Changchun and two at Tientsin. Sandstorms rarely affect the northeast provinces or Korea or places south of Kiangsu and Hupeh, while frequencies are negligible at Harbin, Dairen, Hofei, Nanchang. Kweilin, Chungking, Chengtu and beyond.

However, although the most extreme conditions do not reach to the southern and eastern provinces, poor visibilities there during late winter and spring are evidence of the large amount of dust in the atmosphere. Thus, the percentages of time during each month when visibility at Hong Kong exceeds 17 km are as shown in Table XVIII.

Even offshore from the south China coast, the more vigorous invasions of the northeast monsoon carry considerable dust, and at that time of year the visibility only exceeds 25 km for brief periods associated with the washout of solid impurities by showers. Sandstorms appear to be not infrequent in the mountainous regions of western Szechuan and Changtu Province; Changtu Town and Kantse have 13 and 35 sandstorms each year, most of which occur in winter and spring (Table XVII). The Takla Makan is also noted for its severe sandstorms, though the confining mountain ranges ensure that the storms cannot affect such a large area as those generated in the Gobi and Ordos. Turfan and Kucha have spring maxima and totals of 13 sandstorm days annually while there are 33 days at Hami. However, observations at Urumchi indicate that conditions are rather better to the north of Tien Shan.

Regional classifications of climate

The various classifications

With the general terms used in Köppen's classifications (WARD, 1919), it is possible to recognise some similarities of climate at different places within China and to compare the climate of any particular part of the country with that of other parts of the world. DE MARTONNE (1925) retains some of the advantages of using a system of universal application, but also finds it necessary to introduce some new descriptions to adequately describe the diverse regional climates of China. Thus, his *Mexican* and *Mediterranean Steppe* climates have counterparts elsewhere, but the *Manchurian* and *Chinese* climates have not. SION (1928) and CHU (1929) depart even further from the Köppen tradition; Chu's climatic divisions are of much more restricted application and emphasize such peculiarly local features as KENDREW'S (1957) rainfall regimes and the orographic effects of the Chinling and Nanling Ranges.

Thornthwaite's (1933) well-known system, which reintroduces the concept of universality, produces regional boundaries which differ greatly from those of earlier classifications in some parts of China.

Lu's (1949) classification is described here at considerable length. He employs Köppen's symbols and, with unprecedented attention to detail and taking advantage of the increasing fund of basic climatological information, distinguishes at least 28 separate climatic regions within China.

Chen (1957) follows Chu's (1929) style to a certain extent but, with the aid of Thornthwaite's definitions and much more complete basic data, makes finer distinctions than Chu. For example, Chen separates the *humid* Kwangtung–Fukien coast from *sub-humid* Kwangsi and distinguishes different climatic regions within Taiwan and Hainan.

Köppen (1918)

Köppen's regional classification of the climates of the world is based on temperature, rainfall and seasonal characteristics. Fig.19 (after Ward, 1919, and Chu, 1929) illustrates the large range of climate in Central and East Asia where all Köppen's main climatic divisions A to F are represented.

A *tropical rainy* (A) climate is found in southern Hainan, where Aw'g denotes a *savanna* type with annual rainfall 1,000–1,500 mm, maximum rainfall in autumn, and temperature

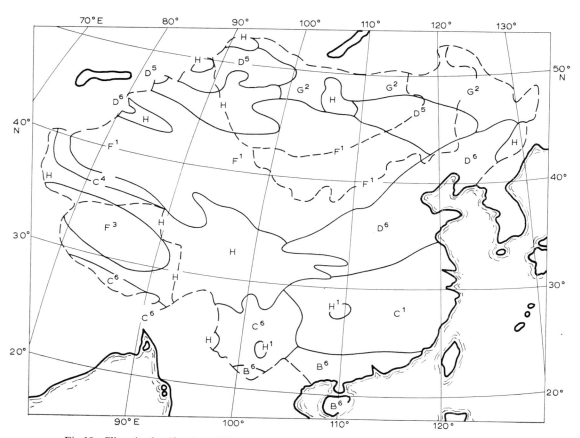

Fig.19. Climatic classification of Köppen. (After Ward, 1919.)

reaching a maximum before the rains. This is not inconsistent with the data for Yulin (see Table LXVIII) where mean annual rainfall is 1,330 mm, maximum rainfall occurs in September–October and highest temperatures are from June to August.

The dry (B) climates stretch in a broad belt through the interior. This region is far from the oceans. During summer it is protected by the Tibetan Plateau from incursions of moist tropical air and during winter it is covered by an intense anticyclone. *Middle-latitude steppe* (BSkw) climates extend from central Inner Mongolia to Dzungaria and cover Kansu, the northernmost parts of Shensi and Shansi and the enclosed lower levels of Chinghai. The *steppes* have annual rainfall averaging 250–500 mm and annual temperature below 18°C. Cold winters constitute the driest season. Paotou with an unreliable rainfall of about 204 mm approximately meets the definition (see Table XXXI). In the west, the *steppes* give way to *deserts* (BW) covering central Sinkiang where mean annual rainfall is mainly less than 100 mm.

Köppen's *warm temperate rainy* (C) climates cover the country south of the Hwang Ho and east of the 1,000 m contour and extend over Taiwan, west Korea and north Hainan. West Korea, east Taiwan and the coastal strip between Shanghai and Foochow are said to have Cfa climate, defined by temperatures over 22°C in the warmest month and with rainfall exceeding 30 mm in the driest winter month. These specifications are met at Mokpo, Taipei, Shanghai, Hangchow and Wenchow (see climatic tables). Most of the low-level country is described as Cw. Mean temperatures in the coldest month should be

Zone	Type[1]	Description
2	Aw′g	Tropical rainy. Savanna. Coldest month's temperature >18°C; maximum temperature before the rains; annual rainfall 1,000–1,500 mm; periodically dry with autumn rainfall maximum.
3	BSkw	Dry winter steppes. Cold winter, mean annual temperature <18°C; warmest month >18°C. Annual rainfall 250–500 mm, and driest season in winter.
4	BW	Desert. Rainfall under 250 mm.
5	Cfa	Warm temperate rainy. Coldest month's temperature > −3°C; warmest month >22°C; driest month's rainfall >30 mm or rainfall of wettest summer month 10 times that of driest winter month.
	Cw	Warm temperate rainy. Coldest month's temperature between −3°C and 18°C; rainfall in wettest summer month at least 10 times that of driest winter month.
8	Df	Sub-arctic. Coldest month's temperature < −3°C; warmest month's temperature > 10°C; no dry season.
9	Dwa	Sub-arctic. Coldest month's temperature < −3°C; warmest month's temperature >22°C; driest in winter.
10	E	Snow. Tundra. Warmest month's temperature <10°C but >0°C.
11	F	Snow. Perpetual frost. Warmest month's mean temperature <0°C.

[1] A = tropical rainy; B = dry; C = warm temperate rainy; D = sub-arctic; E, F = snow climates; S = steppes (rainfall 250–500 mm); W = deserts (rainfall <250 mm); a = temperature of warmest month >22°C; b = temperature of warmest month <22°C, more than 4 months >10°C; c = temperature of only 1–4 months >10°C, coldest month > −36°C; f = constantly moist; g = Ganges temperature maximum before summer rains; k = cold winters, mean annual temperature <18°C, warmest month >18°C; k′ = cold winters, mean annual temperature <18°C, warmest month <18°C; m = monsoon rains; n = frequent fog; s = driest season in summer; s′ = driest season in summer, rainy season in autumn; w = driest season in winter; w′ = driest season in winter, rainy season in autumn; w″ = driest season in winter, rainy season in two parts with short dry season intervening; x = rainfall maximum in late spring or early summer, driest season in late summer.

between 18°C and −3°C, while the rainfall of the wettest summer month should be 10 times that of the driest winter month. Shantung, southern Yunnan, western Taiwan and coastal Kwangtung meet these criteria; but recent statistics indicate that elsewhere the dry season is insufficiently pronounced and that Cf climates are more likely to prevail throughout Anhwei, eastern Hupeh, Kiangsi, Hunan, Kweichow and Kwangsi.

Sub-arctic (D) climates stretch in a broad belt from Heilungkiang and east Korea through the high altitude provinces west of the plain to Szechuan and Changtu Province on the Tibetan Plateau. The temperature of the coldest month is below −3°C and that of the warmest month is above 10°C. Available data confirm that much of this area has Dwa climate with a dry winter and with mean temperature over 22°C in the warmest month. However, CHU (1929) points out that the low-level parts of east Szechuan should more correctly be classed as C than D climates, because winter mean temperatures there are above −3°C. Further to the west much of the Df area might more aptly be classed as Dw because the winters of Changtu Province are very dry.

Snow-climates which are mainly *tundra* (E) are found on the very high country of Chinghai, northwestern Changtu Province, and on the Tien Shan and Kunlun Range which border Sinkiang. In these areas the mean temperature of the warmest month is defined as less than 10°C and more than 0°C, but there are also many places in the Himalayas and in the extreme west where *perpetual frost* (F) prevails.

De Martonne (1925)

De Martonne's division of the Chinese lowlands is consistent with the concept of temperate, subtropical and tropical zones divided by the Nanling and Chinling Ranges (see Fig.20). The division is primarily according to temperature.

To classify the southeast provinces as *Central India monsoon* (B⁶) type conforms with the rainfall and temperature regimes of Kwangtung and Kwangsi. The central provinces and the Yangtze Valley are said to be a distinctive *Chinese* (C¹) type with annual temperature below 20°C and with temperatures in the four warmest months over 10°C. Similarly, the climate which prevails from the North China Plain to west Korea is called *Manchurian* (D⁶) type and is specified by mean annual temperatures below 10°C and a cool rainy summer. Available data indicate that this description applies to Korea and Liaoning but that mean annual temperatures are probably above 10°C further to the south and west.

No distinction is made between the *tundras* and *perpetual frost*. The tops of high mountains and plateaux are classed collectively as a *Himalayan* (H¹) type without warm season, except in Tibet where a *Tibetan* (F³) type has mean annual precipitation below 250 mm. *Aral* (F¹) type climates with mean annual precipitation below 250 mm and a large annual variation of temperature roughly coincide with Köppen's area of *dry* (B) climates but with no distinction between *deserts* and *steppes*.

Sion (1928)

Sion uses three main divisions of *warm*, *temperate* and *cold* in describing the climatic regions of monsoonal Southeast Asia (see Fig.21). *Warm Hong Kong* (B³) type climates are found south of the Chinling Hills and in Taiwan and Hainan. Mean monthly temperatures should exceed 25°C in at least two months and should be below 10°C in four

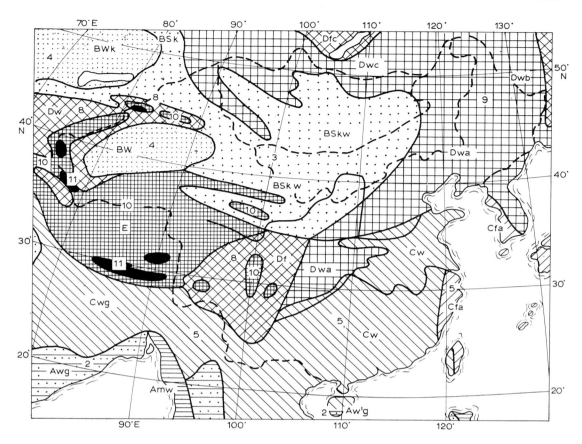

Fig.20. Climatic regions of DE MARTONNE (1925), excluding Korea and Taiwan.
B^6 = central India monsoon type, annual temperatures >20°C; C^1 = Chinese type, annual temperature <20°C, warmest 4 months >10°C; C^4 = Mediterranean steppe type, as C^1, dry summer half year; C^6 = Mexican type, as C^1, small annual temperature range; D^5 = Ukraine type—steppe, annual temperature <10°C, cool summer; D^6 = Manchurian type, as D^5, rainy summer; F^1 = Aral type, annual precipitation <250 mm, large annual temperature range; F^3 = Tibetan type, annual precipitation <250 mm; G^2 = Siberian type, 4 months with mean temperatures >10°C; H^1 = Himalayan type, no warm season.

months or less. This resembles DE MARTONNE'S (1925) *Chinese* type, but makes no special recognition of the generally high temperatures of the southern provinces.

South Yunnan and west Kweichow, which are on the slopes exposed to the summer monsoon, have a *warm Calcutta* (B^4) type climate with an April–May temperature maximum and with copious rains but more than two dry months. Sion (like De Martonne) considers that a "dry" month is one whose annual rainfall in mm is less than double the temperature in °C. Recent data indicate that the temperature maximum in the southern Yunnan–Kweichow region is more towards the middle of the year while Kweiyang has no "dry" months.

A *warm variety temperate* (C^7) climate extends from Shantung westwards along the Hwang Ho, where the mean temperatures during more than four months are below 10°C and during less than four months are below 0°C. The *cold variety temperate* (C^8) climate is defined by a coldest month temperature of less than −2°C. It should probably not extend more than about 150 km southward into Szechuan which has a comparatively

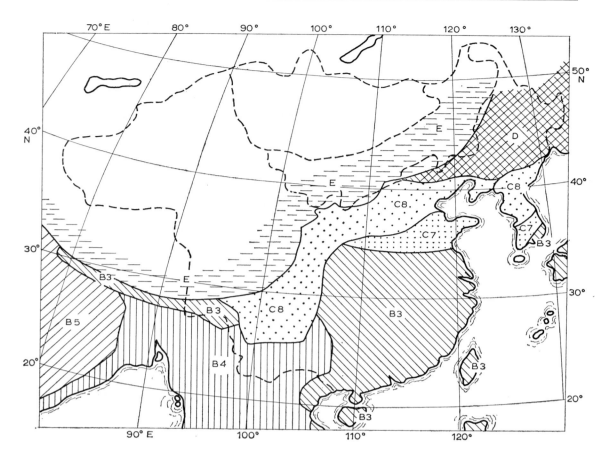

Fig.21. Climatic regions of SION (1928).
B3 = warm climate, Hong Kong type, in 2 or more months temperatures >25°C, in 0–4 months temperatures <10°C; B4 = warm climate, Calcutta, more than 2 dry months, distinct April–May temperature maximum; C7 = temperate climate, warm variety, mean temperatures in more than 4 months <10°C, mean temperatures in 4 months at the most <0°C; C8 = temperate climate, cold variety, as in C7, but temperatures of coldest month < −2°C; D = cold climate, Manchurian type; E = cold climate, high steppe type, annual precipitation <400 mm.

warm winter (cf. data for Chengtu in Table LI and the January mean isotherms in Fig.11).

Data for Pyongyang, Kim Chaek (Joshin), Inchon and Taegu (Taiko) confirm the existence of the three types C8, C7 and B3 in Korea.

Chu (1929)

Chu Co-ching adopts temperature criteria which differ slightly from those of KÖPPEN (1918), DE MARTONNE (1925) and SION (1928) but which achieve similar boundaries in some areas. Chu accepts Sion's *Manchurian* and *steppe* types in the northeastern provinces and Inner Mongolia, but gives recognition to the fact that the climate of east Szechuan is just as temperate as that of the lower Yangtze. He greatly simplifies conditions in the western provinces with only two climatic types—*Tibetan* and *Mongolian* (see Fig.22).

East of the Harbin–Tongking line, Chu stresses the importance of KENDREW'S (1927) three rainfall regions. In *South China* climates the mean annual rainfall is taken to be over 1,000 mm and usually over 1,500 mm, with a maximum between June and September. In the *Central China* or *Yangtze Valley* climate, the mean is over 750 mm, while in the *North China* climate it is 400–600 mm with a maximum in July. In the east, the boundaries separating these three climatic divisions lie along the lines of the Chinling and Nanling Ranges. Thus, they conform with the east–west orientation of mean annual isotherms and isohyets and are practically identical with De Martonne's boundaries. East of the three divisions is considered to have a distinctive "mean annual temperature range" with greatest range in the north and least in the south.

The *South China* type includes Fukien, Kwangtung, Kwangsi, south Hunan and south Kiangsi, where tropical cyclones contribute to the summer rainfall maximum. The mean

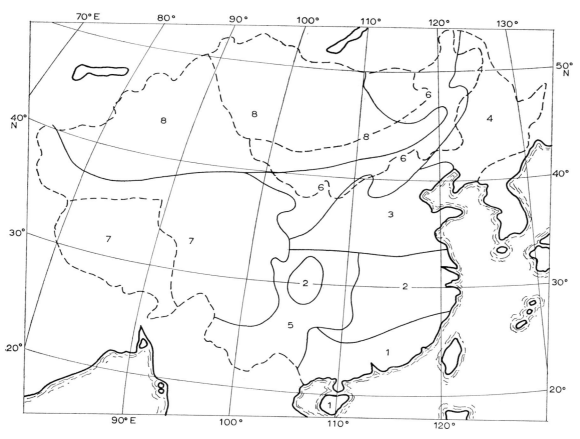

Fig.22. Climatic regions of CHU (1929), excluding Korea and Taiwan.
1 = South China type: coldest month's temperature >10°C, mean annual temperature range 12–20°C, mean annual rainfall >1,000 mm with maximum June–September; 2 = Central China or Yangtze Valley type: mean temperatures <10°C in 4 months at most, mean annual temperature range 18–25°C, rainfall >750 mm; 3 = North China type: mean November temperature <10°C but >0°C, mean annual temperature >10°C, mean annual temperature range 25–35°C, mean annual rainfall 400–750 mm with maximum in July; 4 = Manchurian type: mean annual temperature <10°C, mean temperatures <0°C in 5 or more months, mean annual precipitation 400–600 mm; 5 = Yunnan Plateau type: mean annual temperature 14–18°C, mean annual temperature range 12–15°C, mean annual precipitation mainly >750 mm; 6 = Steppe type: mean annual temperature 5–10°C, mean annual precipitation 200–400 mm; 7 = Tibetan type; 8 = Mongolian Type.

temperature of the coldest month (January) is over 10°C and the region is said to agree with the tropophytic forest region. Lichi, bananas and pineapples are characteristic products and three crops of rice are raised each year.

The *Central China* area covers Chekiang, Anhwei, east Hupeh and Hunan and most of Kiangsi, where at least four winter months have mean temperatures below 10°C. Winter and spring are comparatively wet with *Mai-yü* rains from extratropical depressions during April–June. The *Central* region is the region of the temperate tropophytic forest and much of China's tea is grown there.

The *North China* type is found in Shantung and Honan, north Kiangsu and Anhwei and south Shensi, Shansi and Hopei, where mean annual temperatures are above 10°C and November mean temperatures are between 0°C and 10°C. Maximum rainfall is in July and winters are very dry, while annual rainfall varies greatly from year to year. Wheat and millet are the main crops.

The *Manchurian* type has a mean temperature below 0°C for 5 months and an annual mean below 10°C. Half of the annual rainfall of 400–600 mm falls in July–August. It is the region of the narrow-leaved sclerophyll forest and spring wheat and soy beans are the main crops.

Tu (1936) has proposed refinements to this system of classification which would increase the number of climates from 8 to 24. He divides Chu's *Manchurian* area into three parts consisting of the Hingan-ling Mountains[1], Manchurian Plain and the southeastern mountains. The *Mongolian* type is split into desert, desert-steppe, pasture-steppe, agriculture-steppe and northern mountains, while the *North China* area contains the Great Plain and the Loess Plateau. The *Central China* area is divided into the lower Yangtze Valley, middle Yangtze, the Red Basin and the Gulf of Hangchow, and *South China* into southeast China coast, Sikiang Valley and Hainan. *West China* consists of the Chinling Hills, Sikang (west Szechuan) Mountains and the southwest tableland while the *Tibetan* area is similarly divided into three parts. These proposals form the basis of Lu's (1949) comprehensive classification which is given on pp.51–56.

Thornthwaite (1933)

In the eastern provinces, Thornthwaite's climatic boundaries show the influence on precipitation effectiveness and temperature efficiency of the approximately northeast–southwest orientation of summer isotherms and isohyets (Fig.23). Thus, Thornthwaite's boundaries broadly resemble KöPPEN's (1918) but do not follow CHU's (1929) and DE MARTONNE's (1925) in recognising the effect of the Chinling and Nanling Hills.

Thornthwaite's location of the *arid* and *semi-arid* regions follow general practice, while in the high-altitude southwestern provinces the *tundra* and *taiga* regions correspond approximately with Köppen's *tundra* and *sub-arctic* and with De Martonne's *Tibetan* and *Himalayan* types.

An unusual feature of Thornthwaite's map is the designation of a *semi-arid* climate for Liaoning and the provincial boundary region further to the north. Recent precipitation and temperature data for Dairen indicate that coastal Liaoning might be described more aptly as *subhumid microthermal* (CC[1]).

---

[1] Hingan-ling (Kingan-ling) 46°–53°N 118°–129°E approximately.

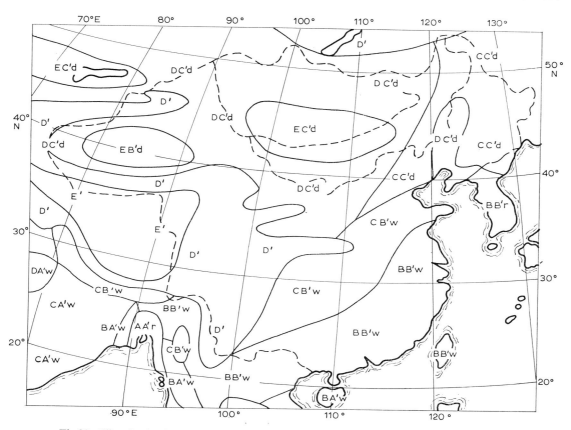

Fig.23. Climatic classification of THORNTHWAITE (1933).

Precipitation effectiveness (P-E index)			Temperature efficiency (T-E index)		
A	128 +	wet	A′	128 +	tropical
B	64–127	humid	B′	64–127	mesothermal
C	32– 63	subhumid	C′	32– 63	microthermal
D	16– 31	semi-arid	D′	16– 31	taiga
E	0– 15	arid	E′	1– 15	tundra
			F′	0	perpetual frost

d = rainfall deficient in all seasons; r = rainfall adequate at all seasons; s = rainfall deficient in summer; w = rainfall deficient in winter.

Elsewhere the climatic divisions appear to be consistent with values of precipitation effectiveness and temperature efficiency calculated from available long-period temperature and rainfall data. Precipitation effectiveness values computed from recent evaporation observations generally confirm Thornthwaite's estimates, but indicate rather more humid conditions at Hong Kong and Chungking.

Lu (1949)

Lu modifies KÖPPEN's (1918) basic principles to provide the greater detail which he considers necessary to distinguish between the diverse climates of China. Köppen's boundaries are defined by the warm-month mean isotherms 0°, 10°, 18° and 22°C, by

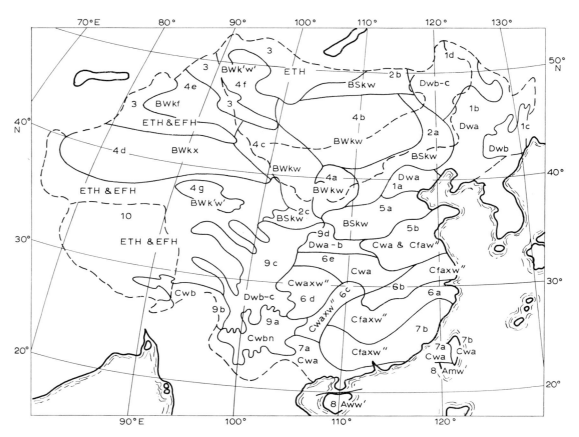

Fig.24. Climatic regions of Lu (1949) excluding Korea.

1 = northeast type, Dw: 1a = Yen-shan, Dwa; 1b = northern lowlands, Dwa; 1c = Changpai-shan, Dwb; 1d = Hingan-ling, Dwb-c.

2. Northern steppes, BSkw: 2a = Inner Mongolia steppes, BSkw; 2c = Chinghai-Kansu steppes, BSkw.

3.Northwest mountain type, ETH and EFH.

4. Northwest desert type, BW: 4a = Ordos, BWkw; 4b = Gobi, BWkw; 4c = A-la Shan, BWkw; 4d = Tarim, Bwkx; 4e = Dsungar, BWkf; 4g = Tsaidam, BWk'w'.

5. North China type, BSkw in north, Cwa or Cfa in south: 5a = Loess Plateau and Haiho Plain, BSkw; 5b = Shantung Hills and Anhwei Plain, Cwa and Cfaw".

6. Central China type, Cfaxw" and Cwaxw": 6a = southern Yangtze Valley, Cfaxw", 6b = northern Yangtze Valley, Cfaxw"; 6c = southwest Hupei and Kweichow Upland, Cwaxw"; 6d = Red Basin, Cwaxw"; 6e = Hanshui Valley, Cwa.

7. South China type, Cwa and Cfaxw": 7a = upper Si Kiang Valley, Cwa; 7b = Nanling Hills, Cfaxw".

8. Hainan type, Aw' and Amw'.

9. West China type, variable, mostly Dwb in north and Cwb in south: 9a = Yunnan Plateau, Cwbn; 9b = southeast Tibet, Cwbg; 9c = east Tibet, Dwb-c; 9d = Chinling, Dwa-b.

10. Tibet type, ETH and EFH.

the cold-month isotherms —3° and 18°C, by the mean annual 18°C isotherm, by the dry month 30 mm isohyet and by the 250, 500, 1,000 and 1,500 mm annual isohyets. These criteria result in large areas of BS, Cw, Cf and Dw each containing large variations of rainfall and temperature, and in climatic boundaries which are divorced from the discontinuities of vegetation and land relief. Lu adheres to Köppen's nomenclature but introduces the January —6° and 6°C isotherms and the 750 and 1,250 mm annual isohyets. Ten main regions and 27 subdivisions are thus described (see Fig.24). The main divisions are the *northeast* type, *northern steppes, northwest mountain, northwest desert, North China, Central China, South China, Hainan, West China* and *Tibet* types.

1. Northeast type (Dw)

Lu fixes the western boundary of the *northeast* (Dw) type climate by annual precipitation of $r = 20 (t + 14)$ mm where t is mean annual temperature in °C. East of this line are coniferous forests, and grains and vegetables flourish; but to the west only arid grasses survive. Lu's Dw climate does not extend very far westward into Mongolia, Liaoning or Hopei, and he partly replaces KÖPPEN's (1918) broad belt of *sub-arctic* (D) climate by *semi-arid steppes* (BS) in Shansi and a *warm temperate* (C) climate in east Szechuan. Four Dw sub-divisions are distinguished.

(*1a*) The Yen-shan[1] (Dwa) area is in central Hopei where January temperature is above −6°C, summer is fairly hot and wet, and annual precipitation is at least 500 mm.

(*1b*) The northern lowlands (Dwa) of Heilungkiang, Kirin and Liaoning have January temperatures below −6°C and even down to −25°C in the north. In winter the soil freezes and rivers are ice-bound, while summer is hot and showery with July temperatures over 22°C. Precipitation varies from 600 mm in the southeast to 400 mm in the northwest.

(*1c*) The Changpai-shan[1] (Dwb) comprises the high country of east Heilungkiang and eastern Kirin. The hottest month has temperatures below 22°C, and orography contributes to an annual rainfall exceeding 750 mm with a maximum in summer. In the low-lying valleys rice is planted in summer while dense coniferous forest covers the higher inaccessible areas.

(*1d*) The Hingan-ling (Dwb-c) covers the extreme northeast of Inner Mongolia where winter is very severe with average January temperatures below −25°C. Average temperature in the hottest month is below 22°C and less than four months have mean temperatures over 10°C. The annual range of temperature is 50°C and precipitation is only 300 mm/year. Agriculture is precarious but coniferous forests thrive.

2. Northern steppes (BSkw)

LU's (1949) *northern steppes* are defined by annual precipitation which is between 20 $(t + 14)$ and 10 $(t + 14)$ mm and which is mainly 200–300 mm. This region is separated from the *North China* region by the −6°C isotherm. Winter is very cold and summer is warm with occasional showers. The *northern steppes* cover only the eastern fringes of KÖPPEN's (1918) *dry winter steppes* but encroach on the northern parts of Liaoning, Hopei, Shansi and Shensi and southern Kansu.

(*2a*) The *Inner Mongolian steppes* (BSkw) cover the eastern edge of the Mongolian Plateau down to north Shansi. July temperatures over 18°C and moderate rainfall enable dry-farming to be undertaken.

(*2b*) The *Chinghai–Kansu steppes* (BSkw) of southern Kansu and the eastern low-level portion of Chinghai are much the same as the Inner Mongolian *steppes* (BSkw) but have warmer winters and slightly cooler summers.

---

[1] Yen-shan: 40°–41°N 118°–119°E approximately; Changpai-shan: 41°–43°N 127°–129°E.

TABLE XIX

COMPARISON OF CLIMATES OF TIEN SHAN AND OF THE NORTHERN TIBETAN BORDER

Northern Tien Shan		Southern Tien Shan		Northern Tibetan border	
height (m)	climate	height (m)	climate	height (m)	climate
1,000	BSKw, BSk'w	1,500	BWkw	1,500	BWkw
1,000–2,000	Dfb	1,500–3,000	BSk'w	1,500–3,000	BSk'w
2,000–3,000	Dfc	3,000–3,500	Dwc	3,000–4,500	Dwb-c
3,000–3,500	ETH	3,500–4,000	ETH	4,500–5,500	ETH
3,500	EFH	4,000	EFH	5,500	EFH

3. Northwest mountain type

This is a general description for the Tien Shan and Altai Ranges where the variations of climate depend mainly on altitude. Lu lists the climates of Tien Shan and compares them with those of the north Tibetan border (see Table XIX).

4. Northwest deserts (BW)

Here annual precipitation is less than 100 mm. The *northwest deserts* coincide with most of KÖPPEN's (1918) *dry winter steppes* and *deserts* and are mainly in inland basins enclosed by hills. Both summer heat and winter cold are intense, and sandstorms are common in spring.

Divisions *4a, b* and *c* in Fig.24 represent the BWkw climatic regions of the Ordos and Gobi and A-la Shan Deserts[1]. Summer showers occur occasionally in the Ordos but less frequently in the Gobi, while average annual rainfall in the Ala-shan Desert is only 40 mm.

(*4a*) The Tarim (BWkx) covers all the province of Sinkiang south of the Tien Shan, together with the low-lying portions of northern Kansu.

(*4b*) Dzungaria (BWkf) has rather more rainfall than the other deserts, and rainfall is fairly well distributed through the year. The winters are very severe, and invasions of cold air in spring are accompanied by sandstorms and duststorms.

(*4c*) Tsaidam[1] (BWk'w') at 2,000–3,000 m altitude to the north of Kunlun Shan has a cool summer and cold winter and maximum rainfall is in late summer and early winter.

5. North China type (BSkw and Cwa or Cfaw'')

This region, which includes most of the Hwang Ho Basin, is separated by the −6°C January isotherm from the *northern steppes* and is bounded by the annual isohyets of 20 $(t + 14)$ mm in the northeast and of 750 mm in the south. Winter is characterised by the duststorms which accompany successive cold fronts. Rainfall is mostly from summer showers and amounts to only 350 mm in the northwest.

(*5a*) The Loess Plateau and Haiho[1] Plain (BSkw) lie north of the Hwang Ho and are described by KÖPPEN (1918) as *sub-arctic* (Dwa) which requires the coldest month's temper-

---

[1] A-la Shan Desert: 39°–41°N 100°–106°E approximately; Tsaidam: 37°N 93°–95°E approximately; Hai Ho (River) enters Po Hai at 39° 00′N 117°42′E.

ature to be below −3°C. In fact, temperatures do not fall this low in the southernmost part of the region.

(5b) The Shantung Hills and Anhwei Plain (Cwa and Cfaw″) have heavier rainfall and higher temperatures.

6. *Central China type (Cfaxw″ and Cwaxw″)*

This climatic region is bounded by the 750 mm annual isohyet in the north, the 22°C July isotherm in the west and the 6° January isotherm in the south. Here the Mai-yü rains of early summer are a prominent feature. Summer is enervating, spring and winter are changeable with the regular passage of Yangtze depressions, while autumn is cool and dry. Cold outbursts from the north bring snow and frost but mean monthly temperatures are never below 0°C. The mean annual precipitation of 1,000–1,500 mm is fairly evenly distributed through the year.

(6a) The southern Yangtze Valley (Cfaxw″), stretching from Chekiang through Kiangsi to Hunan, has rainfall above 1,250 mm. Cold fronts with heavy snowfalls maintain average monthly precipitation of 50 mm throughout the winter. In some parts there are said to be two summer rainfall maxima, one in June from the Mai-yü depressions and the other in August due to typhoons.

(6b) In the northern Yangtze Valley (Cfaxw″), annual precipitation is less than 1250 mm, with drier winters and usually a single summer rainfall maximum.

(6c) Southwest Hupei and Kweichow Upland (Cwaxw″) are regions of generally higher elevation, where rainfall varies from 1,000 mm in the valleys to over 1,500 mm on the more exposed slopes. Rainfall is said to have a June Mai-yü maximum besides that of September, but winter also has much drizzle and fog. According to Lu (1949) most of the summer rain occurs at night.

(6d) In the enclosed valleys of the Red Basin (Cwaxw″) the annual rainfall of less than 1,000 mm occurs mostly in the summer. Summer is very hot and Lu considers that the climate might appropriately be classified as a *South China* type because January temperatures exceed 6°C and snow and frost are rare.

(6e) The Hanshui Valley (Cwa) climate of Hupeh and the valleys to the west has a single summer rainfall maximum and conditions much the same as those of the Red Basin.

7. *South China type (Cwa and Cfaxw″)*

This warm rainy region stretches from south Chekiang across Fukien, Kwangtung and Kwangsi to south Yunnan. The January isotherms of 6°C and 18°C fix its northern and southern limits, and snow and frost are rare. Tropical cyclones contribute to the annual rainfall of 1500–2000 mm.

(7a) In the Upper Si Kiang Valley (Cwa) it is warm throughout the year, with maximum rainfall in July and August.

(7b) In the Nanling Hills (Cfaxw″) area the rainfall maxima are in late spring and August. There is frequent drizzle in winter, and orography contributes to an unusually high winter rainfall on the northeastern coast of Taiwan. Cfb climates occur on hills over 2,000 m.

8. Hainan type (Aw' and Amw')

In these humid tropical areas all monthly temperatures are above 18°C. Tropical cyclones and thunderstorms contribute to the summer rainfall maximum. The island of Hainan and the scattered islets of the South China Sea are described as Aww' and the southwestern coast of Taiwan as Amw'.

9. West China type

Generally the climates of West China are Dwb in the north and Cwb in the south, but patches of BS and BW appear in the leeward valleys and enclosed basins, and ETH or EFH climates cap the peaks. The vertical distribution of climate is mainly as seen in Table XX.

TABLE XX

COMPARISON OF CLIMATES OF CHINLING, EASTERN TIBET, AND NORTHERN YUNNAN AND SOUTHERN TIBET

Chinling		Eastern Tibet		Northern Yunnan and southern Tibet	
height (m)	climate	height (m)	climate	height (m)	climate
1,500	Dwa	1,500	Cwa	1,500	Cwa
1,500–2,500	Dwb	1,500–3,000	Cwb	1,500–4,000	Cwb
2,500–4,000	Dwc	3,000–4,000	Dwb-c	4,000–4,500	Dwb
		4,000–5,500	ETH	4,500–6,000	ETH
		5,500	EFH	6,000	EFH

(*9a*) The Yunnan Plateau (Cwbn), at altitudes of 1,500–3,000 m, has a warm winter and most of the precipitation occurs in summer showers.

(*9b*) The valleys at 3,500–4,000 m on the southeastern Tibetan Plateau (Cwbg) have mild climates with regular frosts, but winter monthly mean temperatures which are not below 0°C. Thunderstorms occur daily during the warm rainy season May–September.

(*9c*) At 3,000–4,000 m on the eastern Tibetan Plateau (Dwb-c), summers are cool with rainfall maxima in June and September, while winters are cold.

(*9d*) The Chinling region (Dwa-b) coincides with part of Köppen's (1918) Dwa region. Summer rainfall is greatest in August and September, while winter is cold and dry.

10. Tibetan (ETH and EFH)

Over these frozen deserts mean monthly temperatures are never greatly above 0°C. However, much of the ground is free from snow and the general snowline is about 6,000 m on the high mountains.

Chen (1957)

Chen determines the rainfall regions in accordance with Thornthwaite's criteria (see Fig.25). The patterns broadly resemble CHU's (1929) but are more complicated due to the

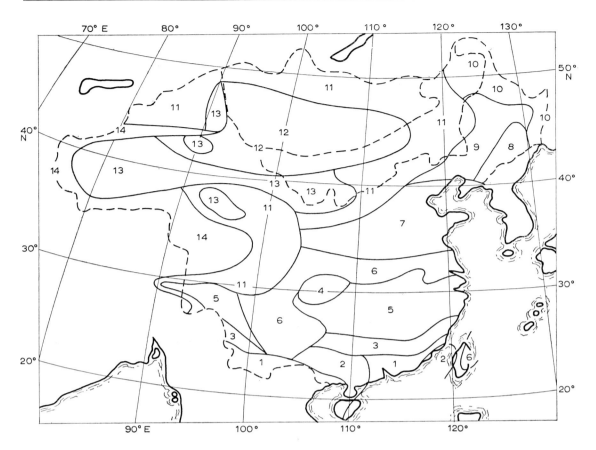

Fig.25. Climatic regions of CHEN (1957), excluding Korea.
1 = tropical humid: mean monthly temperature always >10°C; frosts rare; average rainfall >1,500 mm with double maxima; average annual T-E 114 cm; mositure index 20–100; practically no water deficiency. 2 = tropical subhumid: as in 1, but rainfall mainly 1,200 mm; moisture index 0–20. 3 = subtropical humid: mean temperature of coldest month 8 to 10°C; 10 frost-free months; annual potential evapotranspiration >1,000 mm; rainfall >1,500 mm; moisture index 40–80. 4 = subtropical sub-humid: sheltered by high mountains, warm winters; very hot summers; annual rainfall mainly <1,000 mm. 5 = temperate humid: four distinct seasons; coldest month's temperature 3 to 8°C; rainfall 1,000–1,500 mm with Mai-Yü rains before hot season; fair humid winters; water surplus throughout year. 6 = temperate subhumid: rainfall <1,000 mm. 7 = temperate semi-arid: January −6°C and 0°C isotherms are the region's north and south limits; erratic rainfall with average 500 mm; 75% of rainfall in hot summer. 8 = cold temperate humid: severe winter; cool summer; copious rainfall; snowstorms at start and finish of winter. 9 = cold temperate subhumid: long dry severe winters; rivers frozen 4 months in south, 6 months in north; short warm summer; 60% of the 600 mm rainfall falls in June–August. 10 = frigid subhumid: annual mean temperature <0°C; coldest month's temperature < −23°C large annual temperature range; 8 months winter and only 4 frost-free months. 11 = steppe: scarce erratic precipitation; severe winter and cool summer. 12 = Gobi (frigid arid): extremely cold and dry; very large annual and diurnal temperature range; marginal erratic rainfall <300 mm. 13 = desert (temperate arid): annual rainfall <100 mm. 14 = highland tundra: mostly cold desert with large diurnal temperature range.

increased amount of data which was available by 1957. A distinction is made between the climates of the east and west coasts of Taiwan and Hainan, and Chu's *South China* area is replaced by *tropical humid, tropical sub-humid* and *subtropical humid* zones. Deline-ation of the rainfall regions conforms in some respects with J. H. CHANG's (1955) moisture index patters (Fig.18).

Garnier and Küchler

The specific problem of delimitation of the humid tropics has been approached in different ways by GARNIER (1961) who uses conventional climatological data and by KÜCHLER (1961) who takes the type of vegetation into consideration. According to FOSBERG (1961), Garnier's humid tropical region is a region of maximum human discomfort, while Küchler's is one in which conditions are optimum for plant growth. Garnier's humid tropics include nearly all of Kwangtung and the southernmost parts of Fukien and Kwangsi and are bounded by a line stretching west-southwestwards from Foochow. South of this boundary mean annual rainfall is at least 1,010 mm, and 6–11 months have mean falls of 75 mm or more; 8–11 months have temperatures of 20°C or more and mean relative humidity of at least 65%, while there are 6–11 months when mean vapour-pressure is 20 mbar or more.

KÜCHLER (1961), approaching the problem from a study of the vegetation, selects two

TABLE XXI

SUMMARY OF REGIONAL CLASSIFICATIONS IN FIG.19–25

System	Date	Temperature and rainfall criteria	Divisions in China and Mongolia (east of 100°E longitude), excluding Taiwan	
			number of different climatic descriptions	number of separate climatic areas
KÖPPEN (world classification, see WARD, 1919)	1918	250, 500, 1,000 and 1,500 mm annual isohyets 18°C annual isotherm 0°, 10°, 18° and 22°C isotherms for warmest month −3° and 18° isotherms for coldest month	7	9
DE MARTONNE	1925	250 mm annual isohyet 10° and 20°C annual isotherms 10° isotherm for warmest months	8	12
SION	1928	400 mm annual isohyet 25° isotherm for warmest months −2°, 0° and 10°C isotherms for coldest months	6	6
CHU	1929	200, 400, 600, 750, 1,000 and 1,500 mm annual isohyets 5°, 10°, 14°, 18°C annual isotherms 0° and 10°C isotherms for coldest months	7	9
THORNTHWAITE	1933	Precipitation-effectiveness and temperature-efficiency	8	8
LU	1949	250, 500, 750, 1,000 and 1,250 mm annual isohyets 18°C annual isotherm 0°, 10°, 18° and 22°C isotherms for warmest month −6°, −3°, 6° and 18°C for coldest month	11	23
CHEN	1957	100, 300, 1,000 and 1,500 mm annual isohyets 0° annual isotherm −24°, −6°, 0°, 3°, 8°, 10° isotherms for coldest month	13	15

boundaries. The Si Kiang estuary and Hainan are in a permanently humid region where tropical broadleaf evergreens may be mixed with deciduous trees and patches of grassland. Between this region and a boundary which stretches westward from Foochow is a periodically humid region of deciduous forests comprising most of Kwangtung and Kwangsi and the southern parts of Yunnan and Fukien.

Summary

The main features of the various classifications are summarised in Table XXI. KÖPPEN'S (1918) criteria, which were designed for application on a world scale, distinguish nine different climatic areas east of 110°E longitude. Within these nine areas there are seven types of climate. By the introduction of additional rainfall and temperature criteria, Lu distinguishes 23 separate climatic areas containing 11 different types of climate.

Data sources and studies of particular areas

Data for China, Hong Kong and Macau

Up to this stage China and the neighbouring countries have been treated as a whole, and the broad generalisations owe much to the comprehensive climatological maps and tables published by the Central Weather Bureau (ANONYMOUS, 1953, 1954a,b, 1955, 1960). There are other data sources. Climatological observations for various places in China during the early part of this century are contained in several publications which include the following: data for several stations in the *Annual Reports* (1916–1935) and *Monthly Bulletins* (1946–1947) of Tsingtao Observatory; for a number of customs stations in the *Monthly Meteorological Bulletins* (1928–1936) of the Academia Sinica, National Research Institute of Meteorology, Nanking; for Canton in the *Monthly Meteorological Magazine* and other publications of the Sun Yat Sen Observatory and Canton Meteorological Observatory (1927–1935); for two stations in Hunan during 1932–1938 in a publication of the Meteorological Division of the Cotton Industry Laboratories; for Peking from 1932 onwards in the *Quarterly Meteorological Bulletins* of the National Tsing Hua University Meteorological Observatory; for Wuhan from 1933 onwards in the *Monthly Meteorological Bulletins* of the National Wuhan University; for Wuhu from 1934 in the *Observaciones Meteorológicas* of the Catholic Mission in Wuhu; for some stations in Kwangsi from 1936 onwards in the *Monthly Weather Bulletins* of the Government Weather Bureau in Nanning; for several stations from 1936 onwards in the *Meteorological Bulletins* of Foochow Provincial Observatory.

There are probably many other publications of this nature, and several specialised climatological studies have been published in the *Meteorological Magazine* of the Meteorological Society of China from 1925 onwards. Climatological records for Shanghai commencing in 1874 and records for about 50 Chinese maritime customs stations for shorter periods are contained in various publications of Zikawei Observatory. The relations of climate and cultivation in the province of Kiangsu have been analysed by HU (1947). In Hong Kong, the Royal Observatory has published climatological records and summaries dating back to 1884, together with some detailed accounts of the rainfall of

Hong Kong, Kowloon and the New Territories (STARBUCK, 1950; PETERSON, 1964). Observations for Macau are available in *Resultados das Observações Meteorológicas* from 1952 onwards; some earlier records for Macau were also published, and detailed analyses of the climate are available (CABRITA, 1948; FERREIRA, 1960).

Four areas which for various reasons merit particular attention are Korea, Taiwan, Sinkiang and Szechuan. Korea and Taiwan are treated separately because for many years they have been dissociated from China in the climatological literature. The climate of Szechuan is so different from that of the plains that it also has acquired a special literature. Sinkiang has been included because, while there is a growing interest in the province, existing climatological summaries cover too short an observational period to justify their inclusion in the mean charts.

General accounts of the climates of China are contained in CRESSY's (1934) and KENDREW's (1941, 1957) works. GHERZI (1951) describes the climates of 52 different places in China which are said to possess distinctive "climagraphs". A "climagraph" for any place is constructed by plotting mean temperature as ordinate against mean relative humidity as abscissa for each month of the year. The shape of the polygon formed by joining up the points for successive months is said to give a good representation of climatic characteristics at the place of observation. For example, a large area enclosed by the climagraph indicates large seasonal variations, and a climagraph slanting up to the right draws attention to a warm damp summer. From the climagraphs for 120 stations, Gherzi selects 52 as being sufficiently distinctive to represent the climates of China. Climagraphs for Peking, Wuhan and Hong Kong in Fig.26 emphasize marked differences in the climates of those places. At Wuhan the mean relative humidity is always high though the seasonal range of temperature is nearly as great as that of Peking. Hong Kong has a fairly large seasonal range of humidity but not of temperature.

Korea

Climatological observations were commenced at Pusan, Mokpo, Inchon (Chemulpo), Wonsan and Sunuiju (Yongampo) in 1904, at Kim Chaek (Joshin, Songjin) in 1905, and at Seoul, Pyongyang and Taegu (Taiko) in 1907. The observations for 1904–1911 were published in full by the Japanese Central Meteorological Observatory (*Results of the Meteorological Observations made at the Japanese Meteorological Stations in China and*

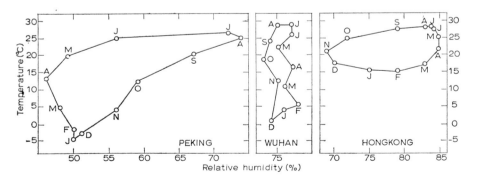

Fig.26. Comparison of mean monthly temperature and relative humidity at Peking, Wuhan and Hong Kong.

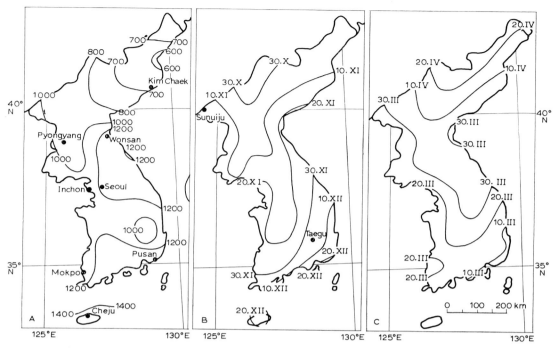

Fig.27. Climatological maps for Korea.
A. Mean annual precipitation (mm).
B. Average first day of snowfall.
C. Average last day of snowfall.

Korea, published yearly), and those from 1911 onwards by the Weather Bureau of Chosen (*Annual Reports of the Weather Bureau of Chosen*). By 1953 the Central Meteorological Office in Seoul had taken over, and in that year information was published for 14 stations but was restricted to places south of 38°N latitude.

In 1962 the Central Meteorological Office in Seoul issued a comprehensive *Climatological Atlas of Korea*. This was based on data for 14 South Korean stations over periods starting at various times between 1931 and 1943 and ending in 1960; North Korea was represented by 11 stations covering periods varying from 7 to 45 years up to 1949. The atlas contains monthly maps for most of the usual climatological elements, and some of the maps dealing with rainfall and temperature are reproduced in Fig.27.

COOK (1964), in a general summary of the climate of Korea, draws attention to the long duration of freezing temperatures. Inland in the north, there are 210 days annually with minimum temperature below freezing. There are 165 such days in the centre of the country and on the north coast, 135 in central coastal areas and 120 in the south. Average minimum temperatures in January are about 2°C on Cheju Do, −2°C on the south coast, −5 to −10°C in central areas, −13°C in the northeast, and −29°C inland near the north boundary where an absolute minimum of −43.6°C has been recorded. In summer, maximum temperatures are above 25°C from the end of April to mid-October in the south, and from early June to early September in the north.

July–September is the typhoon season, and COOK (1964) estimates that South Korea is affected by severe typhoons about once each year and North Korea about once every 6 years. Annual precipitation varies from about 600 mm on the northeast coast to more than 1,200 mm in the south and in central eastern parts of the peninsula.

TABLE XXII

ANNUAL RAINFALL AT SEOUL (AFTER HIRATA, 1917)

Year	Rainfall (mm)
1774	620
1787	2,270
1821	2,580
1876	640
1879	2,290
1887	590
1901	370
1770–1907	1,160 (mean)

McCUNE's (1941) is probably the best known descriptive survey of the climate of Korea. He points out that, while mean January temperatures are generally below freezing and August temperatures are above 22°C, the range is much greater in the north and the interior than in the south and along the coasts. He also draws particular attention to the unreliability of the annual rainfall as illustrated by the totals from ancient unstandardised records made at Seoul as shown in Table XXII.

Using about 20 years' data for 150 observing stations, McCune distinguishes ten different regional climates as shown in Fig.28. Winter temperatures, which govern the type of winter crop and determine whether there are one or two crops, also determine the climatic boundaries. The five southernmost climatic divisions have C climates with mean January temperature above −3°C and winter cropping, while all the rest have D climates. This approximately conforms to Köppen's classifications shown in Fig.19.

McCune describes the climatic regions as follows: The northern interior (*1*) is an isolated mountainous region. It has a long, cold, dry, winter when five months are below 0°C, followed by a short, warm, moist summer. Potatoes, oats and millet are cultivated in the region, and there are forests of firs and pines. The northeastern littoral (*2*) has a cold winter with three months below 0°C and a fairly warm summer. Annual rainfall varies from 720 mm in the central part of this region to double that amount in the south, and fog is common along the coast. Dry crops are produced once a year. The northern west (*3*) is forest-covered hilly land with a very cold, dry winter and mean January temperatures below −8°C. Rainfall varies from 920 mm on the coast to 1,370 mm in the mountains. The warm, wet summer is suitable for single crops of rice and dry grains. In central-west Korea (*4*), the mean January temperatures of −8°C to −6°C are sufficiently high to permit small winter crops of wheat and barley. In the southern west (*5*), January temperatures of −6°C to −3°C and annual rainfall of 860–1,370 mm permit intensive rice production on the alluvial plains and dry crops on the lower mountain slopes, and some winter barley is also grown.

In the south (*6*), the July maximum of rainfall and the secondary peak in April are favourable for irrigated rice cultivation. January mean temperature is below freezing. The mean annual rainfall varies from 890 mm in the east to 1,500 mm in the west, but occasional droughts can be disastrous for the rice crop. Cotton and soybeans are also grown in summer, and barley during the winter. The southeastern littoral (*7*) has a mild winter with January temperatures between −3°C and 0°C and no dry months. Rice is the main crop. The southern littoral (*8*) has a mild winter with mean January temperature above 0°C and a long hot summer. Rainfall is heavy, particularly in the mountains where it exceeds 1,500 mm and the main crops are rice and cotton but dry crops are grown during winter.

Cheju Do (Quelpart Island, *9*) has a warm moist marine climate with January mean temperature of 4°C and annual rainfall of about 1,400 mm, while Ullung Do (Dagelet

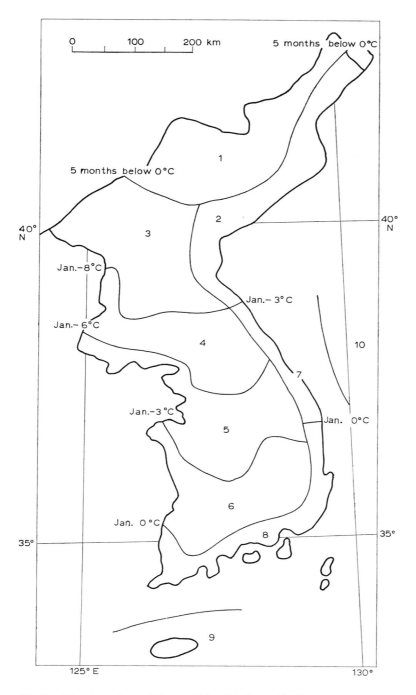

Fig.28. Climatic regions of Korea. (After McCune, 1941.)

Island, *10*) has a mean January temperature of 1°C, very heavy winter rainfall and mean annual rainfall of 1,500 mm.

Taiwan

Climatological observations for a number of stations in Taiwan and the Pescadores are available for as far back as 1896 (ANONYMOUS, 1902), and analyses of the rainfall were published by Taihoku Observatory in 1914 and 1920 (ANONYMOUS, 1914, 1920). Two important factors which combine in determining the climates of Taiwan are the seasonal alternation of monsoon winds and the barrier presented by central and eastern ranges which exceed 3,500 m M.S.L. in many places. Summer rains predominate in the south-west but not in the northeast. Thus, more than 80% of the annual rainfall occurs during April–September at Taichung, Penghu, Ali Shan, Tainan and Hengchun: 60–80% of the annual total occurs in summer at Taipei, Hualien and Taitung but only 26% at Chilung (Keelung). July and August are the wettest months in these places which have summer maxima; but some places also have considerable rainfall in June, which CHI (1964) considers is due to the onset of the summer monsoon combined with the passage of Mai-yü depressions.

C. S. CHEN (1956) writes that, west of the main ranges, the most important climatic boundary is near Tung Shih. South of this point more than 80% of the annual rainfall occurs in summer, while to the north of it winter rains become more important. Close to Tung Shih are the 16°C coldest month isotherm and the isopleth for average 50% sunshine in January. At this boundary also, the tea plantations of the north give way to the cultivation of sugarcane, upland rice, bananas, pineapples and tobacco.

According to C. S. Chen the four main climatic regions in Taiwan are the northern hills, western lowlands, southeast hills and central mountains (see Fig.29). The western lowlands have an area amounting to 30% of the area of the whole island and to 57% of its cultivated area. On these lowlands there is little rainfall in winter and less than 20 mm in November; so cultivation depends considerably on irrigation. One of the chief characteristics of the southeast hills region is its vulnerability to catastrophic floods associated with typhoons.

C. S. Chen's central mountain region covers half of the total area of Taiwan (Fig.29A). It extends to within 35 km of the northern tip of the island and to within 25 km of the southern tip and it reaches the east coast at places. Annual rainfall is mainly over 2,000 mm and increases with altitude to 5,000 mm on the very exposed slopes. July temperatures are below 26°C along the boundaries of this region but temperature falls off at about 0.5° to 0.7°C per 100 m of altitude in the interior. On Tung Shan (Yu Shan), January and July mean temperatures are −0.7°C and 6.8°C respectively. Over most of the high country, frost first appears in mid-October and lasts during 180 days up to mid-April; but on Ali Shan (2,406 m), it has been reported in all months except June, July and August. Natural forest flourishes, with broad-leaf trees at the base of the mountains, mixed forests at mid-levels and coniferous trees at still higher levels. Tea and citrus fruits are grown on the western boundaries of the region.

In C. S. Chen's northern hills region there is a comparatively uniform seasonal distribution of rainfall. Tea and citrus fruits are grown on the slopes, while the level ground produces two crops of paddy rice. The climates of some particular parts of the northern

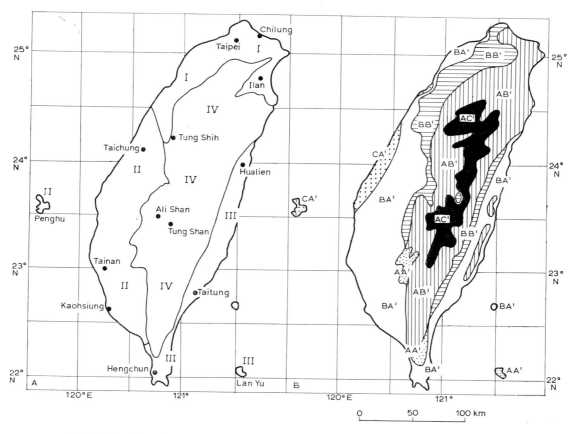

Fig.29. Climatic regions of Taiwan. A. After C. S. CHEN (1956). B. After CHENG and HUANG (1956).

region have been described in detail by C. S. CHEN (1958) and other writers in the *Meteorological Bulletin*. Thus, LIAO and HSU (1961) have analysed the climate of the Tanshui River Basin, and CHANG (1961) has described the climate of Taipei. C. S. CHEN (1958), in writing about the rainfall of Chilung, stresses the abundance and persistence of the rainfall. Chilung has average annual rainfall of 3,159 mm and monthly mean falls ranging from 134 mm in July to 335 mm in December, and during January–March 1908 continuous rain was recorded on 55 consecutive days.

K. T. C. CHENG's (1960) well-known essay on the climates of Taiwan contains a considerable amount of data in tabular form, some of which is reproduced below. Temperature records in Table XXIII for nine stations at low levels show that, although summer mean temperatures are all much the same, winter temperatures vary from about 15°C in the north to 20°C in the south.

K. T. C. Cheng uses mean monthly temperatures at five high-altitude observing stations to describe the lapse-rate and seasonal temperature variation in the mountains. January and July mean temperatures are 10° and 23°C respectively at 840 m, 14° and 23°C at 1,000 m, 6° and 14°C at 2,400 m, 5° and 13°C at 2,700 m, and —3° and 7°C at 3,800 m. The statistics which are reproduced in Table XXIV demonstrate the winter rainfall maximum at Chilung, the early summer Mai-yü maximum at Taichung, the large annual rainfall total for mountainous Ali Shan, and a seasonal variation of rain-days at Taitung which is remarkably small in comparison with the seasonal variation of rainfall there.

TABLE XXIII

MEAN TEMPERATURE AT LOW LEVELS IN TAIWAN

Place	Mean temp.			Absolute maximum	Absolute minimum
	Jan.	July	annual		
Chilung	15.5	28.2	21.9	37.9	5.0
Taipei	15.2	28.2	21.8	38.6	−0.2
Taichung	15.8	27.8	22.4	39.3	−1.0
Tainan	17.1	27.9	22.4	37.8	2.4
Kaohsiung	18.8	28.1	24.4	36.7	7.3
Hengchun	20.5	27.7	24.6	36.2	9.5
Ilan	15.9	27.9	21.9	37.9	4.2
Hualien	17.3	27.4	22.6	36.0	4.4
Taitung	19.0	27.7	23.6	39.5	7.2

Much of the rainfall on the western lowlands and in the mountains is associated with thunderstorms, which occur on an average of 8–10 days a month during June to August and on 35–45 days each year. In the east and northeast and on the small islands, there are only three or four thunderstorm days in each summer month and 10–25 each year. Thunderstorms are generally infrequent in winter, though most places average about one occurrence over the period November–February.

Most places in Taiwan experience prolonged dry spells from time to time (K.T.C. CHENG, 1960) and rainfall is seriously deficient in years when there are few typhoons. However, typhoons also do an immense amount of damage. Gusts up to 65.7m/sec

TABLE XXIV

MEAN MONTHLY AND ANNUAL RAINFALL, RAIN-DAYS AND EVAPORATION IN TAIWAN

	Jan.	Feb.	Mar.	Apr.	May	June	July	Aug.	Sep.	Oct.	Nov.	Dec.	Year
	Rainfall (mm)												
Chilung	319	312	301	222	267	283	134	170	257	255	304	335	3159
Taichung	34	68	101	130	230	388	290	329	144	22	17	27	1780
Taitung	39	42	59	81	163	229	316	303	308	167	89	45	1841
Hengchun	21	28	22	52	173	416	522	539	309	139	61	21	2303
Ali Shan	72	120	172	260	543	798	755	773	489	132	58	87	4259
	Rain-days												
Chilung	22	21	21	18	18	16	10	12	15	19	21	23	215
Taichung	8	10	12	11	13	17	16	17	9	4	5	6	126
Taitung	11	10	13	15	18	14	13	14	15	12	10	10	153
Hengchun	9	8	7	8	12	19	21	22	18	12	9	8	152
Ali Shan	10	11	13	14	21	24	26	25	20	15	9	9	197
	Evaporation (mm)												
Chilung	67	62	80	96	116	140	206	199	160	128	99	75	1427
Taichung	95	90	109	124	151	153	174	163	157	155	120	97	1588
Taitung	113	109	125	137	153	163	185	171	154	152	131	115	1709
Hengchung	154	149	190	197	199	160	160	148	154	182	171	158	2021
Ali Shan	69	67	74	80	78	70	75	69	70	78	75	73	877

have been recorded (K.T.C. CHENG, 1960) and, on average, typhoons account for about 100 lives and destroy 270,000 houses each year (LIAO, 1960).

In a comprehensive study of rainfall and evaporation over Taiwan, CHEN and HUANG (1956) use 10-year mean temperatures and 20-year mean rainfall from 100 stations. Their climatic classification on the Thornthwaite system, which is reproduced in Fig. 29B, shows much more variety than the original Thornthwaite map (Fig.23). CHEN and HUANG'S (1956) classification is said to conform with the distribution of vegetation. Thus, the sugar cane in Taiwan is nearly all confined to their BA' region; the tea-plantations are mostly in the BB' area while the hardwood forests and the tea gardens of northeast Taiwan are in the AB' zone.

Sinkiang

Not very much has been published about the climate of Sinkiang because of lack of organised meteorological observations. However, GHERZI (1951) has described conditions in the province in general terms, while C. Y. CHANG (1949) has written of the climate in relation to topographical features. Chang writes "Sinkiang is bounded on three sides by mighty mountain masses and on the fourth side by wide stretches of barren plateaux.

TABLE XXV

SHORT-PERIOD CLIMATOLOGICAL DATA FOR SINKIANG[1]

Place	Annual rainfall (mm)	Average temperature (°C) Jan.	Jul.	Percentage of possible sunshine	Average number of fog-days per year	Average number of days of sandstorm
Tarim Basin						
Chichioching[2]	51 (6)	−12.8	26.5	75	0	3
Turfan	23 (11)	−10.1	33.4	68	6	13
Hami	42 (6)	−12.0	29.5	71	4	33
Singsingsia	90 (4)	−16.5	21.2	80	8	1
Kara Shahr	57 (12)	−11.2	23.7	70	11	9
Kucha	68 (8)	− 9.6	26.6	68	3	13
Kashgar	119 (10)	− 7.4	26.0	63	1	15
Maral Bashi[2]	55 (2)	− 9.1	25.8	65	1	28
Khotan	56 (3)	− 8.4	24.7	—	3	28
Yarkand	55 (2)	− 8.9	24.3	62	4	23
Charchan	10 (2)	−11.5	24.8	—	5	51
Dzungarian Basin						
Altai	206 (5)	−17.2	21.5	62	16	—
Tsingho	97 (3)	−21.2	25.0	59	5	6
Kuldja	248 (9)	− 9.2	22.6	—	28	7
Shih-ho-tzu[2]	179 (3)	−19.1	24.6	64	13	5
Kitai	183 (3)	−19.1	23.6	64	20	5
Urumchi	276 (7)	−15.8	23.9	59	31	6

[1] For rainfall the period of observation in years is given in parentheses. The means of other elements are for different periods of 2–5 years.
[2] Chichioching (Tsikiotsing): 43°29′N 91°35′E; Maral Bashi (Pa-chu): 39°44′N 78°34′E; Shih-ho-tzu: 44°19′N 86°00′E.

... The province itself is a vast series of inter-mountain basins. The Tien Shan, extending from west to east through the centre, divides it roughly into two natural regions— Dzungaria in the north and the Tarim Basin in the south. Physically and structurally, the Tien Shan also forms a separate region, with a series of inter-mountain troughs, such as the Turfan depression and the Kuldja Plain. By far the largest part of the Tarim Basin is occupied by a huge desert of drifting sand dunes, generally known as the Takla Makan and one of the most formidable deserts in the world. No rivers flow from Sinkiang to the sea except the Black Irtysh which drains northward into the Ob and thus into the Arctic Sea. The Akosu, Kizil and Konche Rivers combine to form the Tarim. At the foot of the mountains, between the drifting sand dunes and the hard boulder fans, lie the oases, ribbons of intensive cultivation and close settlement following the rivers and branching canals. The Tarim Basin is one the driest areas of the world."

While remarking that rainfall is less in the open desert, C. Y. Chang gives short-period annual rainfall averages at Kucha, Khotan and Charkhlik as 76, 41 and 5 mm respectively. A more recent series of observations around the edges of the Tarim, which also unfortunately covers an inadequate period, is shown in Table XXV (ANONYMOUS, 1960). These totals are of much the same order as C. Y. Chang's.

Table XXV indicates that slightly more rainfall is recorded around the edges of the Dzungaria Basin and particularly in the enclosed valleys of Tien Shan such as the Kuldja and Urumchi. Monthly mean values indicate that several places have a spring rainfall maximum and a few places have a maximum in early summer; but there is no common pattern, and Altai has maxima in both December–January and in June (see also p.25). Table XXV indicates the large seasonal temperature variation of 35° to 45°C over the province, the low winter temperatures in Dzungaria and the comparatively mild winters of the Kuldja Valley.

Szechuan

P. K. CHANG (1941) discusses the climate in relation to topography. His work was based mainly on observations over a period of three to four years and, at the time of publication, Szechuan's boundary with the old province of Sikang (or more recently Changtu) was along 102°E longitude.

The Red Basin is bounded in the northwest by the Min and Chiunghsia Ranges at 3,000–4,000 m. To the west and southwest the Tasueh and Taliang Shan attain heights in excess of 1,500 m. To the northeast are the Tapa and Mitsang Mountains and to the southeast is the Fangtou Range at 1,000–2,500 m. The Yangtze Gorge lies between the Tapa and Fangtou Ranges.

The Red Basin is at 300–700 m M.S.L. and extends over 200,000 km² which is half the area of the whole province. Chang draws attention to a few important features of the climate. First, in the eastern part of the basin there are rainfall maxima in both May–June and September–October. This is confirmed by the records for a longer period given for Chungking (see Table LIV). Secondly, in the western part of the basin a marked maximum of rainfall associated with thunderstorms in July–August accounts for 50% of the annual rainfall, and this is not inconsistent with records for Chengtu (see Table LI). Thirdly, the May–June rainfall in the east exceeds that in the west (see Table LI and LIV).

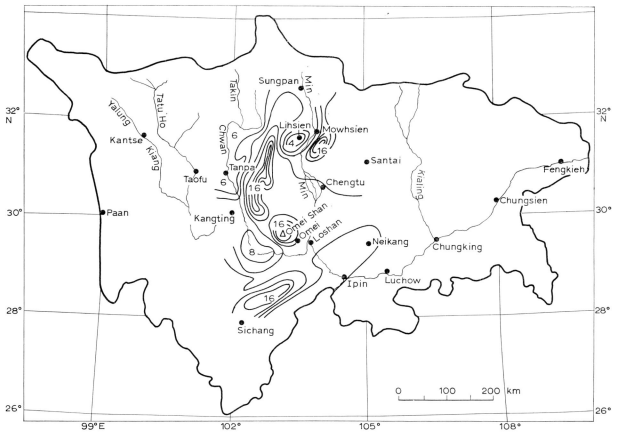

Fig.30. The province of Szechuan and annual rainfall (hundreds of mm) in the mountains. (After Lu, 1947.)

As the ranges surrounding the Red Basin protect it from many of the cold outbursts from the north, winter mean temperatures in the basin are 4°–6°C higher than at places further eastward along the Yangtze. However, winters in the basin are still colder than in places of the same latitude elsewhere in the world. Mean summer temperatures do not differ greatly from those recorded further eastwards along the Yangtze, but Chungking is noted for its hot summer afternoons and an extreme temperature of 44°C which was recorded in August.

P. K. CHANG (1941) distinguishes nine separate climatic regions in Szechuan and describes their topography, climates and characteristic vegetation with fine distinction. Five of the regions cover the Red Basin and the east Szechuan gorge area and have Cwan climates (in Köppen's classification). In the southeast border region, climates are Cwan and CWb, while in the southwest at the foot of Mount Tahliang they also include Dwb and Dwc. Cwan, Cwb and Dwb climates are said to exist in the Tapa Mountains and the northeast region.

P. K. Chang's northwest region is on the edge of the Tibetan Plateau and extends to within 300 km of the present Szechuan–Changtu border. In this region Dwb and Dws climates occur in forest belts at 3,600–4,000 m M.S.L., while cold deserts (ETH) exist above 4,000 m. Steppe climates (BS) are found on many of the slopes and in valleys where sheltering reduces rainfall. Thus at Lihsien (Lifan) and Mowhsien, which lie in the Min

Valley between Mounts Chiuting (4,710 m M.S.L.) and Chiunglai (4,500–5,000 m M.S.L.), mean rainfall assessed over a short period amounted to 262 and 447 mm annually.

GHERZI'S work (1951) contains a descriptive account of the seasonal changes at Ipin (Suifu) and describes five different seasonal regimes of temperature and humidity within the Szechuan Basin as follows (see Fig.31): The southeast and gorge region represented by observations at Chungking, Chunghsien (Chungchow) and Fengkieh (Kweichow); the Chengtu Plain represented by observations at Chengtu and Santai (Tungchwan); the valleys of the northwest border by Kangting (Tatsienlu) and Mowhsien, the southwest border by Sichang (Ningyuan) and the southern basin by Ipin.

Fig.31 demonstrates the low temperatures maintained throughout the year at Kangting (2,558 m) compared with those at lower levels, the persistently high relative humidity near the river at Ipin and Chungking and the large seasonal variation of humidity at Sichang.

HANSON-LOWE (1941) describes such a variety of temperature contrasts and of vegetation types among the peaks and deep gorges of the "northwest border" that any simple general description of the climate there might appear impracticable. However, LU (1947) has made a comprehensive study of the rainfall on the Changtu–Szechuan border between 95° and 105°E longitudes and has published a rainfall map for the Min and Takin Chwan Valleys. Rainfall is greatest on the exposed eastern slopes and least in the valleys, and there are tremendous variations from the arid bottoms of the more sheltered gorges to the forest-covered high slopes. Lu demonstrates the effect of orography at various places near Omei Shan as shown in Table XXVI.

A modified version of Lu's rainfall map for the Min and Tatu Valleys is shown in Fig.30. On those slopes near the edge of the Tibetan Plateau which are exposed to the summer southeasterlies, annual precipitation is usually about 2,000 mm with a record total of 9,236 mm recorded at Omei Shan during the 13 months of August 1932–August

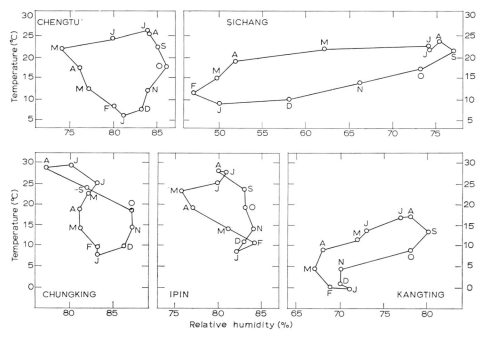

Fig.31. Climates within the Szechuan Basin.

TABLE XXVI

EFFECT OF OROGRAPHY IN THE ENVIRONMENT OF OMEI SHAN

Station	Distance from Omei Shan (km)	Height (m)	Mean annual rainfall (mm)
Omei Shan	0	3,097	1,860
Omei	15	800	1,510
Loshan	33	320	1,220
Neikiang	160	352	1,060

1933 of the second polar year. To the west of the main ridges, annual rainfall is about 600 mm and the aridity of the upper Min and Tatu Valleys is evident from average annual falls of 450 mm and 650 mm at Tanpa and Sungpan.

Lu states that, further westward beyond Tasueh Shan, rainfall decreases generally and varies from 700 to 800 mm on the high slopes to less than 500 mm in the valleys. Thus, Changtu at 3,200 m M.S.L. has an average annual rainfall of 553 mm.

Acknowledgements

The author is greatly indebted to Mr. P. C. Chin of the Hong Kong Royal Observatory for his cooperation in the selection of material and for his advice on both scientific and linguistic questions, and to Messrs. S. Tse and T. S. Li for their assistance in extracting data.

References

ALGUE, J., 1904. *The Cyclones of the Far East*. Philippine Weather Bur., Manila, 283 pp.

ANONYMOUS, 1902. *Meteorological Observations in Formosa, 1896–1901*. Taihoku Meteorol. Obs., Taipei, 138 pp.

ANONYMOUS, 1914. *The Climate, Typhoons and Earthquakes of the Island of Formosa*. Taihoku Meteorol. Obs., Taipei, 80 pp.

ANONYMOUS, 1920. *The Rainfall in the Island of Formosa*. Taihoku Meteorol. Obs., Taipei, 139 pp.

ANONYMOUS, 1937. *Weather in the Chinese Seas and in the Western Part of the North Pacific Ocean*. Meteorol. Office, London, 2: 771 pp.

ANONYMOUS. 1945. *Monthly Meteorological Charts of the Western Pacific Ocean*. Meteorol. Office, London, 120 pp.

ANONYMOUS, 1953. *Climatic Atlas for China—Central Weather Bur., Publ. (Peking)*, 501: 64 pp. (Chinese).

ANONYMOUS, 1954a. *Rainfall Data*. Geophysics Sect. Chinese Sci. Res. Inst., Peking, 723 pp. (Chinese).

ANONYMOUS, 1954b. *Temperature Data*. Geophysics Sect. Chinese Sci. Res. Inst., Peking, 862 pp. (Chinese).

ANONYMOUS, 1955. *Climatic Atlas for China—Central Weather Bur., Publ. (Peking)*, 502: 55 pp. (Chinese)

ANONYMOUS, 1956. World distribution of thunderstorm days. Part 2: Tables of marine data and world maps. *W.M.O. Tech. Publ.*, 21: 71 pp.

ANONYMOUS, 1960. *An Atlas of Chinese Climatology*. Map Publ. Soc., Peking, 293 pp. (Chinese).

ANONYMOUS, 1961. *Climatological and Oceanographic Atlas for Mariners*. U.S. Weather Bur.–U.S. Hydrograph. Office, Washington, D.C., 2: 159 pp.

ANONYMOUS, 1962a. *Climatic Atlas of Korea*. Central Meteorol. Office, Seoul, 299 pp.

ANONYMOUS, 1962b. *Stream Drift Charts of the World*. U.S. Naval Oceanog. Office, Washington, D.C., 12 charts.

ANONYMOUS, 1963. Hong Kong meteorological records. *Hong Kong Roy. Obs., Tech. Mem., Suppl.,* 5: 19 pp.

ANONYMOUS, 1965. *Power Spectrum Analysis of Typhoon Winds.* Royal Observatory, Hong Kong, unpublished.

BARTHOLOMEW, J., 1958. *Times Atlas of the World.* Times Publishing Company, London, 1: 34 pp., 24 plates.

BELL, G. J., 1964. *Investigation of Hong Kong Rainfall.* Royal Observatory, Hong Kong, unpublished.

BIOT, E., 1840. Recherches sur la température ancienne de la Chine. *J. Asiatique,* 10: 530 pp.

BLANFORD, H. F., 1876. The meteorology and climate of Yarkand and Kasgar (Sinkiang). *Mem. Indian Meteorol. Dept.,* 1: 35–111.

BRUZON, E. et CARTON, P., 1930. *Le Climat de l'Indochine et les Typhons de la Mer de Chine.* Imprimerie d'Extrême-Orient, Hanoi, 310 pp.

CABRITA, M. T. F., 1948. The climate of Macau. *Serv. Meteorol. Macau, Sci. Note,* 5: 7 pp.

CHANG, CHIH-YI, 1949. Land utilization and settlement possibilities in Sinkiang. *Geograph. Rev.,* 39: 57–75.

CHANG, JEN-HU, 1955. Climate of China according to the new Thornthwaite classification. *Ann. Assoc. Am. Geographers,* 45: 393–403.

CHANG, KEE-KAR, 1958. The relation between long-period fluctuations of the frequency of occurrence of typhoons in the Pacific and long-period fluctuations of the general circulation. *Acta Meteorol. Sinica,* 29: 135–138.

CHANG, PAO-KUN, 1934. On the duration of the four seasons in China. *Collected Sci. Papers (Meteorology) Acad. Sinica,* 1954: 273–323.

CHANG, PAO-KUN, 1941. Climatic regions of the Szechuan Province. *Collected Sci. Papers (Meteorology) Acad. Sinica,* 1954: 393–439.

CHANG, YEUCH-NGO, 1961. The climate of Taipei. *Meteorol. Bull. (Taipei),* 7: 15–24 (Chinese, with English sum.).

CHEN, CHENG-SIANG, 1956. The geographical regions of Taiwan. *Fu-Min Geograph. Inst. Econ. Develop., Taipei,* 1: 7–14 (Chinese, with English sum.).

CHEN, CHENG-SIANG, 1957. The climatic regions of China. *Meteorol. Bull. (Taipei),* 3: 1–9 (Chinese with English sum.).

CHEN, CHENG-SIANG, 1958. Rainfall of the Port Keelung. *Meteorol. Bull. (Taipei),* 4: 1–5 (Chinese, with English sum.).

CHEN, CHENG-SIANG and HUANG, TSUNG-HUI, 1956. *The Climatic Classification and Moisture Belts in Taiwan.* Fu-Min Geograph. Inst. Econ. Develop., Taipei, 28 pp. (Chinese, with English sum.).

CHENG, CHWEN-SHU, 1949. *Chinese Synoptic Weather Patterns.* Shanghai Obs., Shanghai, 116 pp.

CHENG, K. T. C., 1960. Le climat du Taiwan. *Meteorol. Bull. (Taipei),* 6: 1–10. (Chinese, avec résumé Franç.).

CHENG, KUO, 1963. China's meteorological services today. *Weather,* 18: 366–372.

CHEVALIER, S., 1895. Essay on the winter storms on the coast of China. *Shanghai Meteorol. Soc., Ann. Rept.,* 3: 48 pp.

CHI, CHI-HSUN, 1964. Plum rains in Taiwan. *Taiwan Prov. Weather Bur., Meteorol. Bull.,* 10: 1–12 (Chinese, with English sum.).

CHIN, P. C., 1958. Tropical cyclones in the western Pacific and China Sea area from 1884 to 1953. *Roy. Obs. Hong Kong, Tech. Mem.,* 7: 84 charts.

CHU, CO-CHING, 1929. Climatic provinces of China. *Natl. Res. Inst. Nanking, Meteorol. Mem.,* 1929: 11 pp.

CHU, CO-CHING, 1934. Southeast monsoon and rainfall in China. *Collected Sci. Papers (Meteorology) Acad. Sinica,* 1954: 475–493.

CHU, CO-CHING and YUAN, M. H., 1963. *Phenology.* General Science Publications, Peking, 107 pp.

COOK, CHAEPYO, 1964. *Korean Weather Service.* Central Meteorological Office, Seoul, 41 pp.

CRESSY, G. B., 1934. *China's Geographic Foundations.* McGraw-Hill, New York, N.Y., 436 pp.

DE MARTONNE, E., 1925. Traité de géographie physique. In: *Types de Climats,* Paris, pp.229–261.

FABER, S. E. and BELL, G. J., 1963. *Typhoons in Hong Kong and Building Design.* Engineering Society of Hong Kong, Hong Kong, 29 pp.

FERREIRA, H. A., 1960. O clima de Portugal, provincia de Macau. *Serv. Meteorol. Nacl. (Lisbon),* 10: 40 pp. (Portuguese).

FOSBERG, F. R., 1961. A project for mapping the humid tropics. *Geograph. Rev.,* 51: 333–338.

FRITSCHE, H., 1878. The climate of eastern Asia. *J. N. China Branch Roy. Asiatic Soc.,* 12: 127–335.

FROC, A., 1900. The atmosphere in the Far East during the six cold months. *Shanghai Meteorol. Soc., Ann. Rept.,* 7: 90 pp.

FROC, A., 1901. The atmosphere in the Far East. *Shanghai Meteorol. Soc., Ann. Rept.,* 8: 85 pp.

FROC, A., 1920. *Atlas of the Tracks of 620 Typhoons, 1893–1918*. Zikawei Observatory, Shanghai, 40 pp.

GARNIER, B. J., 1961. Mapping the humid tropics: climatic criteria. *Geograph. Rev.*, 51: 339–346.

GHERZI, E., 1928. *Étude sur la Pluie en Chine, 1873–1925*. Zikawei Observatory, Shanghai, 1: 83 pp.; 2: 198 pp.

GHERZI, E., 1930–1941. *Typhoons in 1928, etc.* Zikawei Observatory, Shanghai, 26–60 pp. annually.

GHERZI, E., 1944. *Climatological Atlas of East Asia*. Zikawei Observatory, Shanghai, 175 pp.

GHERZI, E., 1951. *The Meteorology of China*. Serv. Meteorol., Macau, 1: 423 pp.; 2: 23 Tables.

HANSON-LOWE, J., 1941. Notes of the climate of the South Chinese–Tibetan borderland. *Geograph. Rev.*, 31: 444–453

HEYWOOD, G. S. P., 1950. Hong Kong typhoons. *Roy. Obs., Hong Kong, Tech. Mem.*, 3: 23 pp.

HEYWOOD, G. S. P., 1954. The pressure of typhoon winds on structures. *Proc. UNESCO Symp. Typhoons, Tokyo, 1954*, pp.11–15.

HU, HUAN-YONG, 1947. A geographical sketch of Kiangsu Province. *Geograph. Rev.*, 37: 609–617.

HUNG, K. R., 1951. Fogs at Waglan Island and their relationship to fogs in Hong Kong harbour. *Hong Kong Roy. Obs., Tech. Note*, 3: 8 pp.

JEFFRIES, C. W., 1938. *Meteorological Results, 1937, Appendix II*. Royal Observatory, Hong Kong, 7 pp.

JORDAN, C. L. and FORTNER, L. E., 1960. Estimation of surface wind speeds in tropical cyclones. *Bull. Am. Meteorol. Soc.*, 41: 9–13.

KAO, Y. S., 1948. General circulation of the lower atmosphere over the Far East. *Mem. Inst. Meteorol., Acad. Sinica*, 16: 1–7.

KAO, Y. H. and TSANG, Y. Y., 1957. *Typhoon Tracks and some Related Statistical Analyses*. Scientific Publishers, Peking, 136 pp.

KENDREW, W. G., 1941. *The Climates of the Continents*. Oxford Univ. Press, London, 473 pp.

KENDREW, W. G. 1957. *Climatology*. Oxford Univ. Press, London, 400 pp.

KOO, CHEN-CHIAO, 1951. The importance and influence of the Tibetan Plateau on the general circulation over East Asia. *Acta Meteorol. Sinica*, 22: 283–303.

KÖPPEN, W., 1918. Klassifikation der Klimate nach Temperatur, Niederschlag und Jahresverlauf. *Petermann's Geograph. Mitt.* 64: 193–203, 243–248.

KÜCHLER, A. W., 1961. Mapping the humid tropics: vegetation criteria. *Geograph. Rev.*, 51: 346–347.

LAHEY, J. F., BRYSON, R. A. and WAHL, E. W., 1958. *Atlas of Five-day Normal Sea-level Pressure Charts for the Northern Hemisphere*. Univ. Wisconsin Press, Madison, Wisc., 78 pp.

LEE, J., 1936. Precipitation on the islands along the Chinese coast. *Monthly Weather Rev.*, 64: 287–291.

LIAO, SHYUE-YIH, 1960. A disussion of meteorological diasters in Taiwan. *Meteorol. Bull., (Taipei)*, 6: 1–29 (Chinese, with English sum.).

LIAO, SHYUE-YIH and HSU, CHIN-HUAI, 1961. Climate over the drainage basin of Tanshui River of northern Taiwan. *Meteorol. Bull. (Taipei)*, 7: 1–6.

LU, A., 1937. The cold waves of China. *Collected Sci. Papers (Meteorology) Acad. Sinica, Peking*, 1954: 137–161.

LU, A., 1944. Chinese climatology. *Collected Sci. Papers (Meteorology) Acad. Sinica, Peking*, 1954: 441–466.

LU, A., 1947. Precipitation in the south Chinese–Tibetan borderland. *Geograph. Rev.*, 37: 89–93.

LU, A., 1949. The climatic provinces of China. *Collected Sci. Papers (Meteorology) Acad. Sinica, Peking*, 1954: 467–474.

LUI, S. R., 1957. Climate of Lhasa. *Monthly Weather (Peking)*, 1957: 15–19.

McCUNE, S., 1941. Climatic regions of Korea and their economy. *Geograph. Rev.*, 31: 95–99.

MIN TIEH, T., 1941. Soil erosion in China. *Geograph. Rev.*, 31: 570–590.

NEEDHAM, J. and WANG LING, 1959. *Science and Civilisation*. Cambridge Univ. Press, London, 3: 877 pp.

PAULUS, J. L. H., 1965. Indian Ocean and Taiwan rainfalls set new records. *Monthly Weather Rev.*, 93: 331–335.

PEACOCK, J. E., 1952. Hong Kong meteorological records. *Hong Kong Roy. Obs., Tech. Mem.*, 5: 45 pp.

PETERSON, P., 1964. The rainfall of Hong Kong. *Hong Kong Roy. Obs., Tech. Note*, 17: 22 pp.

RAMAGE, C. S., 1952. Diurnal variation of summer rainfall over east China, Korea and Japan. *J. Meteorol.*, 9: 83–86.

RAMAGE, C. S., 1954. Non-frontal crachin and the cool season clouds of the China Seas. *Bull. Am. Meteorol. Soc.*, 35: 404–411.

SION, J., 1928. Asie des moussons, classification des climats. *Géographie Universelle*, 9: 15–19.

SPENCER, J. E., 1941. Chinese place names and the appreciation of their geographic realities. *Geograph. Rev.*, 31: 79–94.

STARBUCK, L., 1950. A statistical survey of Hong Kong rainfall. *Hong Kong Roy. Obs., Tech. Mem.*, 2: 14 pp.

STARBUCK, L., 1951. A statistical survey of typhoons and tropical depressions in the western Pacific and China Sea area, 1884–1947. *Hong Kong Roy. Obs., Tech. Mem.*, 4: 41 charts.

TANNEHILL, I. R., 1927. Some inundations attending tropical cyclones. *U.S. Monthly Weather Rev.*, 55: 453–456.

TAO, SHIH-YEN, 1948. The mean surface air circulation over China. *Collected Sci. Papers (Meteorology) Acad. Sinica, Peking*, 1954: 575–583.

THOMPSON, B. W., 1951. An essay on the general circulation of the atmosphere over southeast Asia and the west Pacific. *Quart. J. Roy. Meteorol. Soc.*, 77: 569–597.

THORNTHWAITE, C. W., 1933. The climates of the earth. *Geograph. Rev.*, 23: 433–440.

THORNTHWAITE, C. W., 1948. An approach toward a rational classification of climate. *Geograph. Rev.*, 38: 54–94.

THORP, J., 1936. *Geography of the Soils of China*. Nat. Geol. Surv. China, Nanking.

TU, CHANG-WANG, 1936. Climatic provinces of China. *Meteorol. Mag.*, (Nanking), 12: 487–518.

TU, CHANG-WANG, 1938. The air masses of China. *Collected Sci. Papers (Meteorology) Acad. Sinica, Peking*, 1954: 163–202.

TU, CHANG-WANG and HWANG, SZE-SUNG, 1945. The advance and retreat of the summer monsoon in China. *Bull. Am. Meteorol. Soc.*, 26: 9–22.

WARD, R. DE C., 1919. A new classification of climates. Geograph. Rev., 8: 188–191.

WATTS, I. E. M., 1959. The effect of meteorological conditions on tide height at Hong Kong. *Hong Kong Roy. Obs., Tech. Mem.*, 8: 30 pp.

WATTS, I. E. M., 1962. The diurnal variation of frequency of precipitation over southeast Asia. *Reg. Conf. Southeast Asian Geographers, Kuala Lumpur*, 1962.

YAO, CHEN-SHENG, 1946. The stationary cold fronts of central China and the wave disturbances developed over the Lake Basin. *Natl. Res. Inst., Nanking, Meteorol. Mem.*, 20 pp.

YEH, T. C., 1955. The influence of the Tibetan Plateau upon the general circulation of the atmosphere and the weather conditions in China. *J. Sci.*, 6: 29-33.

Annual Reports. Central Meteorol. Office, Seoul. Since 1953.

Annual Reports of the Weather Bureau. Since 1911. Weather Bur. Chosen (Tyosen), Zinsen.

Meteorological Results, Surface Observations. Since 1884. Royal Obs., Hong Kong.

Results of the Meteorological Observations made at the Japanese Meteorological Stations in China and Korea. Since 1904. Central Meteorol. Obs., Tokyo.

TABLE XXVII

CLIMATIC TABLE FOR HARBIN
Latitude 45°45′N, longitude 126°38′E, elevation 143.1 m

Month	Mean sta. press. (mbar)	Mean daily temp.[1] (°C)	Mean daily temp. range (°C)	Temp. extremes		Relative humidity %	Mean precip.[1] (mm)	Max. precip. (24 h)	Max. snow cover (cm)
				highest (°C)	lowest (°C)				
Jan.	1005.9	−20.1	11.8	1.4	−41.4	79	4.3	5.2	7.5
Feb.	1003.5	−15.8	13.5	12.3	−39.9	76	5.6	10.5	14.6
Mar.	999.9	− 6.0	12.7	19.3	−29.1	66	16.7	16.8	17.0
Apr.	993.7	5.8	12.8	29.6	−14.0	59	23.3	30.7	14.0
May	989.7	14.0	13.2	35.6	− 3.7	58	43.9	62.0	0.0
June	986.5	19.8	12.2	39.0	3.8	70	92.0	78.6	0.0
July	987.4	23.3	10.5	39.1	9.6	77	166.6	146.5	0.0
Aug.	989.5	21.6	10.3	37.8	6.0	78	119.1	113.4	0.0
Sept.	995.0	14.3	11.9	32.0	− 3.0	75	52.2	54.3	0.0
Oct.	999.0	5.7	12.3	23.1	−14.8	66	36.4	27.6	4.0
Nov.	1002.2	− 6.6	10.7	18.8	−31.6	71	12.4	15.2	10.0
Dec.	1004.5	−16.7	10.8	6.1	−35.6	78	4.9	7.7	7.0
Annual	996.3	3.3	11.9	39.1	−41.4	71	577.4	146.5	17.0

Month	Mean evap. (mm)	Number of days			Number of cloudy days	Mean sunshine (h)	Most freq. wind direction	Mean wind speed (m/sec)	Number of days sand-storm
		precip.	thunder-storm	fog					
Jan.	9.0	5.4	0.0	1.6	2.0	197.3	W	4.5	0.0
Feb.	33.5	4.7	0.0	1.9	2.3	218.3	S	4.5	0.3
Mar.	84.0	6.0	0.0	1.4	4.5	246.8	W	5.2	0.7
Apr.	232.5	6.9	0.9	0.6	8.3	234.4	W	6.0	0.3
May	243.5	10.9	3.0	0.5	8.5	256.6	SW	5.6	0.0
June	253.5	14.9	6.3	0.7	12.8	257.7	S	4.6	0.0
July	214.0	15.5	9.1	1.9	11.8	266.7	S	4.2	0.0
Aug.	163.0	13.0	3.1	1.3	11.5	253.5	S	3.9	0.0
Sept.	167.5	12.0	4.3	2.4	6.0	216.8	S	4.3	0.0
Oct.	119.5	7.0	0.9	1.5	4.8	213.7	S	5.1	0.3
Nov.	51.0	6.3	0.0	1.2	3.3	181.5	SSW	5.2	0.0
Dec.	13.5	5.6	0.0	1.3	2.3	176.8	S	4.6	0.0
Annual	1584.5	108.2	27.6	15.4	77.8	2715.1	S	4.8	1.7

[1] Mean daily temperature: 1909–1942, 1949–1952; mean precipitation: 1909–1942, 1946–1952.

TABLE XXVIII

CLIMATIC TABLE FOR CHANGCHUN
Latitude 43°52′N, longitude 125°20′E, elevation 215.7 m

Month	M.S.L. press. (mbar)	Mean daily temp.[1] (°C)	Mean daily temp. range (°C)	Temp. extremes		Relative humidity (%)	Mean precip.[1] (mm)	Max. precip. (24 h)	Max. snow cover (cm)
				highest (°C)	lowest (°C)				
Jan.	1027.3	−16.8	12.1	3.7	−35.7	72	5.9	7.7	24.5
Feb.	1024.5	−12.7	13.2	13.7	−36.0	69	5.9	12.8	24.0
Mar.	1019.3	− 3.9	12.5	19.9	−27.7	59	15.3	23.8	23.2
Apr.	1012.3	6.6	13.5	30.2	−15.5	52	22.5	21.1	7.2
May	1007.4	14.4	13.7	34.3	− 3.4	54	52.1	40.2	Trace
June	1003.5	20.0	12.4	39.5	4.4	66	110.2	86.1	0.0
July	1003.8	23.5	10.3	38.3	9.0	76	171.9	97.0	0.0
Aug.	1006.6	21.9	10.6	37.0	6.4	78	139.3	107.9	0.0
Sept.	1012.7	14.9	12.8	33.1	− 2.7	72	54.3	79.1	1.0
Oct.	1018.1	6.8	13.0	29.8	−13.4	66	33.0	35.8	14.0
Nov.	1022.3	− 4.2	11.1	24.0	−29.4	66	16.2	60.4	27.0
Dec.	1025.7	−13.6	11.1	7.5	−34.7	71	5.3	18.1	22.0
Annual	1015.4	4.7	12.2	39.5	−36.0	67	631.9	107.9	27.0

Month	Mean evap. (mm)	Number of days			Number of cloudy days	Mean sunshine (h)	Most freq. wind direction	Mean wind speed (m/sec)	Number of days sand-storm
		precip.	thunder-storm	fog					
Jan.	29.0	5.6	0.0	0.5	2.6	201.2	WSW	3.4	0.0
Feb.	56.5	4.3	0.0	0.5	2.6	207.8	WSW	3.5	0.7
Mar.	102.0	5.7	0.1	0.3	5.5	248.6	WSW	4.2	1.0
Apr.	165.5	7.1	0.4	0.3	7.8	243.4	WSW	4.8	1.0
May	318.5	11.1	1.8	0.1	10.0	258.8	WSW	4.5	0.0
June	310.0	14.6	6.4	0.7	10.2	263.5	SW	3.4	0.0
July	212.0	15.8	6.9	0.8	12.8	247.4	SSW	2.8	0.0
Aug.	173.0	13.0	3.3	1.2	9.8	249.1	SSW	2.4	0.0
Sept.	173.5	10.2	1.9	0.7	6.5	266.6	S	2.8	0.0
Oct.	132.0	8.0	0.9	0.5	5.5	221.1	WSW	3.4	0.0
Nov.	75.5	5.8	0.1	0.5	4.8	179.2	WSW	3.8	0.0
Dec.	26.5	5.0	0.0	0.3	2.7	185.9	WSW	3.6	0.3
Annual	1865.0	106.7	21.9	6.4	80.8	2745.0	WSW	3.6	3.0

[1] Mean daily temperature: 1909–1942, 1947–1952; mean precipitation: 1903–1942, 1947–1952.

TABLE XXIX

CLIMATIC TABLE FOR URUMCHI
Latitude 43°47′N, longitude 87°37′E, elevation 912.6 m

Month	Mean sta. press. (mbar)	Mean daily temp.[1] (°C)	Mean daily temp. range (°C)	Temp. extremes		Relative humidity (%)	Mean precip.[1] (mm)	Max. precip. (24 h)	Max. snow cover (cm)
				highest (°C)	lowest (°C)				
Jan.		−15.8	12.3	3.3	−34.4	87	8.5	8.9	31.0
Feb.		−13.6	10.8	8.1	−41.5	89	15.4	12.0	28.0
Mar.		− 4.0	9.9	19.0	−33.4	80	15.1	28.2	25.0
Apr.		8.5	12.3	28.0	−12.0	52	32.9	17.0	16.0
May		17.7	13.2	35.7	− 9.7	44	24.9	27.6	0.0
June	no	21.5	12.0	36.7	4.4	49	32.8	45.7	0.0
July	data	23.9	12.5	38.1	7.5	46	16.3	10.6	0.0
Aug.		21.9	12.5	38.0	5.5	44	34.9	33.8	0.0
Sept.		16.7	13.1	34.9	1.5	46	14.7	16.3	0.0
Oct.		6.1	10.0	27.9	−18.3	62	47.4	36.6	6.3
Nov.		− 6.2	9.4	19.4	−36.6	84	22.2	12.6	32.7
Dec.		−13.0	11.0	6.5	−36.3	86	11.2	8.8	32.0
Annual		5.3	11.6	38.1	−41.5	65	276.3	45.7	32.7

Month	Mean evap. (mm)	Number of days			Number of cloudy days	Mean sunshine (h)	Most freq. wind direction	Mean wind speed (m/sec)	Number of days sand- storm
		precip.	thunder- storm	fog					
Jan.	10.5	9.7	0.0	5.6	4.6	186.5	SW	1.1	0.0
Feb.	16.0	11.2	0.0	6.0	8.0	147.5	N	1.5	0.0
Mar.	53.0	6.8	0.0	2.6	5.4	193.7	N	1.6	0.0
Apr.	228.0	7.8	0.2	0.6	6.2	203.2	NW	2.6	0.3
May	330.5	5.3	1.0	0.6	4.4	296.9	NW	2.9	1.7
June	364.5	5.8	0.6	0.0	6.8	270.7	NW	2.3	0.7
July	359.5	6.8	2.2	0.0	6.4	279.0	NW	2.0	0.7
Aug.	349.0	4.8	0.8	0.2	5.2	268.2	NW	2.3	0.3
Sept.	250.5	4.0	0.2	0.2	2.2	262.7	NW	2.3	1.3
Oct.	118.0	8.8	0.0	1.4	6.8	212.8	NW	2.2	1.3
Nov.	36.5	11.5	0.0	5.0	8.2	143.7	NW	1.6	0.0
Dec.	10.5	11.0	0.0	3.3	7.0	145.8	SW	1.6	0.0
Annual	2126.5	93.7	5.0	31.0	71.2	2606.9	NW	2.0	6.3

[1] Mean daily temperature: 1951–1953; mean precipitation: 1907–1911, 1930–1931.

TABLE XXX

CLIMATIC TABLE FOR SHENYANG
Latitude 41°46′N, longitude 123°26′E, elevation 416 m

Month	Mean sta. press. (mbar)	Mean daily temp.[1] (°C)	Mean daily temp. range (°C)	Temp. extremes		Relative humidity (%)	Mean precip.[1] (mm)	Max. precip. (24 h)	Max. snow cover (cm)
				highest (°C)	lowest (°C)				
Jan.	1028.3	−12.8	12.3	8.8	−33.1	68	5.6	13.4	20.9
Feb.	1025.7	− 9.1	13.6	14.3	−32.7	64	6.3	15.0	21.0
Mar.	1020.7	− 0.8	11.5	20.2	−25.0	57	13.9	50.3	18.3
Apr.	1014.3	8.8	13.3	30.3	− 9.6	54	29.7	41.4	5.0
May	1009.0	16.2	13.1	33.9	− 2.0	58	65.5	65.6	0.0
June	1004.7	21.6	12.4	39.3	6.5	66	96.2	86.5	0.0
July	1004.5	24.9	9.9	38.9	10.7	76	176.7	119.4	0.0
Aug.	1007.0	23.6	10.0	38.3	9.6	78	161.6	148.7	0.0
Sept.	1013.7	17.0	12.6	33.9	− 1.0	73	73.7	137.0	0.0
Oct.	1019.3	9.4	12.8	30.4	−10.0	70	40.9	52.9	14.7
Nov.	1023.8	− 0.9	10.9	25.2	−26.3	65	23.1	37.4	25.6
Dec.	1027.0	− 9.8	9.6	11.2	−32.2	67	9.6	17.0	21.4
Annual	1016.6	7.3	11.9	39.3	−33.1	66	710.7	148.7	25.6

Month	Mean evap. (mm)	Number of days			Number of cloudy days	Mean sunshine (h)	Most freq. wind direction	Mean wind speed (m/sec)	Number of days sand-storm
		precip.	thunder-storm	fog					
Jan.	26.0	3.9	0.0	4.2	2.8	190.3	N	2.3	0.0
Feb.	69.5	3.4	0.0	2.8	3.0	200.0	N	2.6	0.7
Mar.	113.5	5.5	0.5	2.0	5.4	237.5	SSW	3.3	1.0
Apr.	261.0	6.2	0.4	2.4	7.0	244.8	SSW	3.7	0.3
May	310.0	10.2	2.1	0.6	8.4	237.5	SSW	3.4	0.0
June	232.5	12.3	4.6	0.8	9.5	257.3	SSW	2.8	0.3
July	214.0	14.7	3.9	1.2	12.6	229.8	S	2.4	0.0
Aug.	153.0	12.5	3.1	2.6	9.5	232.3	S	2.0	0.0
Sept.	162.0	8.8	2.5	4.0	5.9	235.1	S	2.1	0.0
Oct.	119.5	6.9	1.1	5.2	4.6	223.8	S	2.6	0.0
Nov.	68.5	5.2	0.1	4.0	4.2	179.5	N	2.7	0.0
Dec.	28.5	4.0	0.0	3.8	3.1	172.7	N	2.5	0.3
Annual	1758.0	93.3	17.8	33.6	75.8	2660.4	S	2.7	2.7

[1] Mean daily temperature: 1905–1942, 1947–1952; mean precipitation: 1935–1943, 1948–1953.

TABLE XXXI

CLIMATIC TABLE FOR PAOTOU
Latitude 40°34′N, longitude 109°50′E, elevation 1044.2 m

Month	Mean sta. press.[1] (mbar)	Mean daily temp.[1] (°C)	Mean daily temp. range (°C)	Temp. extremes		Relative humidity (%)	Mean precip.[1] (mm)	Max. precip. (24 h)	Max. snow cover (cm)
				highest (°C)	lowest (°C)				
Jan.		−11.8	14.7	6.4	−29.6	63	0.6	1.2	5.0
Feb.		− 9.1	14.4	13.4	−26.9	64	3.6	3.4	8.0
Mar.		− 0.1	14.7	20.7	−22.2	49	4.3	4.3	6.0
Apr.		7.2	15.1	26.4	−11.8	48	19.4	20.1	0.0
May		15.5	15.6	33.4	− 2.4	49	31.4	29.4	0.0
June	no	20.9	15.1	36.8	3.7	52	28.5	39.6	0.0
July	data	22.7	13.1	38.4	9.2	64	81.8	41.5	0.0
Aug.		20.7	12.8	35.5	4.9	69	75.9	43.0	0.0
Sept.		14.5	15.4	31.7	− 2.3	61	30.8	32.6	0.0
Oct.		8.0	15.7	25.6	− 9.9	59	23.0	24.0	0.0
Nov.		− 1.9	13.0	15.5	−22.0	61	4.0	6.2	4.0
Dec.		−10.0	14.3	6.7	−32.8	61	0.6	1.2	4.0
Annual		6.4	14.5	38.4	−32.8	58	204.2	43.0	8.0

Month	Mean evap. (mm)	Number of days			Number of cloudy days	Mean sunshine (h)	Most freq. wind direction	Mean wind speed (m/sec)	Number of days sand-storm
		precip.	thunder-storm	fog					
Jan.		1.2	0.0	0.6	1.0	227.1	N	2.0	1.3
Feb.		3.6	0.0	0.6	3.8	209.3	E	2.4	2.0
Mar.		3.0	0.0	0.4	3.5	257.7	E	2.5	3.0
Apr.		4.6	1.2	0.0	3.5	277.1	W	3.5	5.0
May		5.6	4.2	0.4	4.8	284.9	E	2.5	3.0
June	no	7.2	5.8	0.2	5.5	288.8	E	2.8	1.3
July	data	10.2	11.4	0.0	5.3	281.3	SE	2.3	1.7
Aug.		9.8	5.6	0.6	7.5	253.5	ESE	2.2	0.0
Sept.		5.8	3.6	0.6	4.0	271.4	ESE	2.1	1.0
Oct.		4.0	1.4	0.0	3.8	251.1	ESE	2.2	0.0
Nov.		3.4	0.0	1.0	4.5	184.3	ESE	2.3	0.3
Dec.		1.6	0.0	0.0	1.0	209.1	NW	1.9	0.3
Annual		60.0	33.2	4.4	48.0	2986.8	ESE	2.5	19.0

[1] Mean daily temperature: 1935–1937, 1949–1953; mean precipitation: 1935–1937, 1940–1952.

TABLE XXXII

CLIMATIC TABLE FOR PEKING
Latitude 39°57′N, longitude 116°19′E, elevation 52.3 m

Month	M.S.L. press.[1] (mbar)	Mean daily temp.[1] (°C)	Mean daily temp. range (°C)	Temp. extremes		Relative humidity (%)	Mean precip.[1] (mm)	Max. precip. (24 h)	Max. snow cover (cm)
				highest (°C)	lowest (°C)				
Jan.	1025.8	− 4.7	11.4	14.2	−22.8	50	3.7	13.7	23.0
Feb.	1024.3	− 1.9	11.7	18.5	−17.8	50	4.6	10.7	8.5
Mar.	1019.5	4.8	12.7	28.1	−13.8	48	8.4	24.5	16.0
Apr.	1013.5	13.7	13.9	35.8	− 3.3	46	17.2	67.1	0.0
May	1008.2	20.1	13.8	38.1	3.4	49	34.6	109.2	0.0
June	1003.7	24.7	13.2	42.6	10.1	56	77.9	202.6	0.0
July	1002.5	26.1	10.2	40.5	14.9	72	242.6	224.7	0.0
Aug.	1006.1	24.9	10.0	38.3	11.3	74	140.9	144.0	0.0
Sept.	1012.7	19.9	12.1	34.3	1.6	67	58.2	96.2	0.0
Oct.	1017.4	12.8	13.8	31.1	− 4.7	59	15.9	36.6	0.0
Nov.	1022.2	3.8	11.5	24.2	−13.5	56	10.5	54.8	11.0
Dec.	1024.6	− 2.7	10.9	13.5	−19.6	51	2.9	10.8	12.5
Annual	1015.0	11.8	12.1	42.6	−22.8	56	623.1	224.7	23.0

Month	Mean evap. (mm)	Number of days			Number of cloudy days	Mean sunshine (h)	Most freq. wind direction	Mean wind speed (m/sec)	Number of days sandstorm
		precip.	thunderstorm	fog					
Jan.	50.5	2.5	0.0	1.4	3.0	205.5	NNW	2.3	0.4
Feb.	90.0	2.5	0.0	3.2	4.6	197.2	N	2.6	0.3
Mar.	128.0	3.0	0.0	1.6	5.4	236.6	N	2.9	0.9
Apr.	249.0	3.7	0.6	0.7	6.8	239.0	SSW, NNW	3.3	0.7
May	302.5	5.8	2.8	0.4	7.6	265.1	SSW	2.9	0.6
June	283.0	8.1	6.0	2.2	9.2	261.2	N	2.3	0.2
July	217.5	13.3	8.3	2.1	12.7	219.8	E	1.7	0.1
Aug.	147.0	10.8	5.1	2.0	11.2	223.7	E	1.6	0.0
Sept.	144.0	6.6	2.9	1.4	7.2	228.9	W	1.9	0.0
Oct.	95.0	3.3	0.6	1.6	4.9	244.1	E	2.2	0.0
Nov.	70.0	2.5	0.0	2.2	4.5	193.4	E	2.3	0.1
Dec.	52.0	1.6	0.0	2.4	3.7	192.3	NW	2.5	0.3
Annual	1828.5	63.7	26.5	21.2	80.8	2704.6	E	2.4	3.6

[1] Mean daily temperature: 1915–1937, 1940–1952; mean precipitation: 1841–1952.

TABLE XXXIII

CLIMATIC TABLE FOR WONSAN[1]
Latitude 39°11′N, longitude 127°26′E, elevation 35.1 m

Month	M.S.L. press. (mbar)	Mean daily temp. (°C)	Mean daily temp. range (°C)	Numb. of days with max. temp. (<0°C)	Numb. of days with min. temp. (<0°C)	Relative humidity (%)	Mean precip. (mm)	Number of days with snow cover	Max. snow cover (cm)
Jan.	1023.2	− 3.8	9.1	13	30	53	28.5	18	57.3
Feb.	1022.0	− 2.4	9.1	7	27	56	31.5	16	76.3
Mar.	1019.4	2.5	9.0	1	21	59	46.0	9	54.9
Apr.	1014.8	9.7	10.6	–	2	62	67.6	1	26.4
May	1010.8	14.9	10.6	–	–	68	86.4	–	–
June	1007.6	19.0	8.9	–	–	77	125.7	–	–
July	1007.6	22.7	7.3	–	–	83	272.9	–	–
Aug.	1008.7	23.4	7.3	–	–	83	312.1	–	–
Sept.	1013.8	18.8	9.0	–	–	67	177.6	–	–
Oct.	1018.8	13.1	10.5	–	–	66	69.7	–	–
Nov.	1021.5	5.8	9.3	–	11	59	61.1	2	18.1
Dec.	1022.7	− 0.9	8.7	6	27	52	28.6	10	48.0
Annual	1015.9	10.2	9.1	27	118	66	1307.7	56	76.3

Month	Soil temp. at 50 cm (°C)	Number of days with rain (⩾ 1.0 mm)	Number of clear days	Fog	Number of cloudy days	Mean sunshine (h)	Number of days with rain (⩾ 10 mm)	Mean max. temp. (°C)	Dew point temp. (°C)
Jan.	−0.2	3	15		5	204.5		1.0	−10.7
Feb.	−0.5	4	13		5	204.3		2.4	− 9.0
Mar.	1.5	5	10		8	231.8		7.4	− 4.4
Apr.	8.8	5	6		10	235.4		15.4	2.1
May	15.1	7	5		12	235.8		20.7	8.1
June	20.0	9	3	no	16	207.1	no	23.9	14.3
July	23.3	14	2	data	19	172.7	data	26.8	19.3
Aug.	24.7	13	3		17	176.4		27.5	20.1
Sept.	21.1	9	6		12	198.2		23.7	14.5
Oct.	15.6	5	11		6	222.9		18.7	7.0
Nov.	8.8	5	12		5	190.5		10.7	− 0.9
Dec.	2.9	3	15		4	191.4		3.6	− 8.6
Annual	11.8	82	101		119	2471.0		15.1	7.5

[1] Period: 1904–1949.

TABLE XXXIV

CLIMATIC TABLE FOR TIENTSIN
Latitude 39°06′N, longitude 117°10′E, elevation 3.3 m

Month	Mean sta. press. (mbar)	Mean daily temp.[1] (°C)	Mean daily temp. range (°C)	Temp. extremes		Relative humidity (%)	Mean precip.[1] (mm)	Max. precip. (24 h)	Max. snow cover (cm)
				highest (°C)	lowest (°C)				
Jan.	1028.1	− 4.1	9.9	12.1	−20.4	61	3.0	16.3	3.3
Feb.	1025.7	− 1.8	10.4	18.1	−18.6	58	3.6	12.6	3.0
Mar.	1021.4	4.8	11.8	28.8	−14.0	55	8.5	34.9	5.0
Apr.	1014.7	13.4	13.0	35.7	− 2.5	52	16.6	46.1	0.0
May	1009.0	19.8	13.6	41.4	3.6	54	28.1	48.3	0.0
June	1004.1	24.5	12.5	42.7	9.4	62	62.7	78.6	0.0
July	1003.5	26.8	9.8	42.9	16.2	75	176.5	130.4	0.0
Aug.	1006.2	25.9	8.9	39.6	13.7	76	149.2	163.4	0.0
Sept.	1014.1	21.0	10.2	35.6	4.7	69	45.2	58.6	0.0
Oct.	1019.5	14.1	11.3	32.4	− 2.5	63	15.4	44.9	0.0
Nov.	1024.5	4.9	9.8	25.6	−12.3	61	12.8	75.4	3.0
Dec.	1027.1	− 1.9	9.0	14.2	−18.8	59	5.1	18.8	1.0
Annual	1016.5	12.8	10.8	42.9	−20.4	62	526.7	163.4	5.0

Month	Mean evap. (mm)	Number of days			Number of cloudy days	Mean sunshine (h)	Most freq. wind direction	Mean wind speed (m/sec)	Number of days sand-storm
		precip.	thunder-storm	fog					
Jan.	45.9	2.2	0.0	3.5	5.8	184.1	SW	2.3	0.1
Feb.	70.1	2.3	0.0	3.0	6.8	188.8	E	2.5	0.3
Mar.	119.6	2.9	0.1	2.0	6.0	232.6	SSE	3.0	0.6
Apr.	222.7	3.5	0.5	0.8	7.5	245.1	SW	3.3	1.0
May	274.6	5.5	1.7	0.5	9.3	284.4	SSE	3.1	0.4
June	266.4	8.5	4.8	0.7	8.5	272.5	SE	2.7	0.1
July	236.1	13.0	5.1	1.7	13.5	253.4	SSW	2.2	0.0
Aug.	162.1	11.3	3.2	1.0	10.0	231.6	SW, SSW	1.8	0.0
Sept.	166.3	6.1	2.1	0.8	8.3	232.5	SW	2.1	0.0
Oct.	134.8	3.5	0.4	2.5	6.0	241.3	SSW	2.3	0.0
Nov.	103.5	2.6	0.0	4.7	4.0	190.5	NNE	2.4	0.0
Dec.	43.9	2.3	0.0	5.3	5.3	176.3	SSW	2.4	0.3
Annual	1846.0	63.8	17.9	26.5	90.8	2723.2	SSW	2.5	2.8

[1] Mean daily temperature: 1904–1938, 1946–1952; mean precipitation: 1904–1952.

TABLE XXXV

CLIMATIC TABLE FOR PYONGYANG[1]
Latitude 39°01′N, longitude 125°49′E, elevation 27.4 m

Month	M.S.L. press. (mbar)	Mean daily temp. (°C)	Mean daily temp. range (°C)	Numb. of days with max. temp. (<0°C)	min. temp. (<0°C)	Relative humidity (%)	Mean precip. (mm)	Number of days with snow cover	Max. snow cover (cm)
Jan.	1026.5	− 8.1	10.6	22	31	74	14.7	16	24.6
Feb.	1024.4	− 4.8	10.5	12	28	70	11.4	7	26.0
Mar.	1020.5	1.7	10.4	2	25	66	25.5	2	15.2
Apr.	1015.7	9.5	12.4	–	5	63	46.0	–	10.0
May	1011.2	15.5	12.2	–	–	66	66.7	–	–
June	1007.1	20.6	11.6	–	–	71	75.7	–	–
July	1007.0	24.2	8.8	–	–	80	237.3	–	–
Aug.	1007.9	24.4	8.8	–	–	80	228.1	–	–
Sept.	1013.8	18.9	10.9	–	–	75	112.4	–	–
Oct.	1019.6	11.9	12.3	–	2	73	45.1	–	0.4
Nov.	1023.0	3.4	10.2	2	19	73	40.8	2	19.3
Dec.	1025.8	− 4.8	9.7	15	30	74	20.9	10	17.6
Annual	1016.9	9.4	10.7	53	140	72	924.6	37	26.0

Month	Soil temp. at 50 cm (°C)	Number of days with rain (≥ 1.0 mm)	Number of clear days	Fog	Number of cloudy days	Mean sunshine (h)	Number of days with rain (≥ 10 mm)	Mean max. temp. (°C)	Dew point temp. (°C)
Jan.	− 0.3	3	13		5	200.9		−2.7	−10.7
Feb.	− 0.6	3	12		6	207.1		0.6	− 8.6
Mar.	1.2	4	9		8	238.3		7.2	− 3.6
Apr.	8.6	5	7		9	249.5		16.0	2.7
May	15.2	7	5		11	266.3		22.0	8.9
June	20.4	7	3	no	12	261.2	no	26.9	14.7
July	23.9	12	2	data	17	209.8	data	29.1	20.2
Aug.	25.3	10	3		15	212.9		29.3	20.4
Sept.	22.0	7	6		9	234.7		24.8	14.0
Oct.	16.1	6	10		5	242.6		18.5	7.1
Nov.	9.1	7	9		6	181.1		8.8	− 0.2
Dec.	3.1	4	11		6	180.5		0.1	− 7.4
Annual	12.0	75	90		109	2684.9		15.1	8.1

[1] Period: 1907–1949.

TABLE XXXVI

CLIMATIC TABLE FOR DAIREN
Latitude 38°54′N, longitude 121°38′E, elevation 95.6 m

Month	M.S.L. press. (mbar)	Mean daily temp.[1] (°C)	Mean daily temp. range (°C)	Temp. extremes highest (°C)	Temp. extremes lowest (°C)	Relative humidity (%)	Mean precip.[1] (mm)	Max. precip. (24 h)	Max. snow cover (cm)
Jan.	1027.9	− 5.2	7.8	10.6	−19.9	63	9.7	35.7	17.5
Feb.	1025.8	− 3.6	7.5	14.0	−19.3	62	7.4	30.2	38.0
Mar.	1021.7	1.9	7.8	18.8	−14.8	80	14.6	47.1	11.0
Apr.	1015.7	9.4	9.0	28.3	− 4.0	57	25.1	48.0	2.5
May	1010.6	15.5	9.1	32.4	1.4	61	45.6	55.2	0.0
June	1006.2	20.3	8.4	35.0	10.2	71	48.9	83.4	0.0
July	1005.4	23.7	6.6	36.1	14.9	83	155.4	189.6	0.0
Aug.	1007.3	24.5	6.6	35.7	14.8	80	132.9	161.5	0.0
Sept.	1014.2	19.9	7.8	31.6	5.8	70	66.1	104.4	0.0
Oct.	1020.1	13.8	8.3	27.7	− 2.9	64	32.5	70.3	1.0
Nov.	1023.9	5.3	8.1	23.0	−11.6	62	25.0	46.6	23.0
Dec.	1026.7	− 1.9	7.5	13.0	−18.9	61	13.9	50.6	17.0
Annual	1017.1	10.3	7.9	36.1	−19.9	66	575.4	189.6	38.0

Month	Mean evap. (mm)	Number of days precip.	Number of days thunder-storm	Number of days fog	Number of cloudy days	Mean sunshine (h)	Most freq. wind direction	Mean wind speed (m/sec)	Number of days sand-storm
Jan.	42.0	3.9	0.0	1.0	4.0	193.2	NNW	4.8	0.0
Feb.	79.5	3.2	0.0	1.4	4.5	198.8	NNW	4.8	0.0
Mar.	117.0	3.9	0.1	2.7	6.2	242.6	S	5.3	0.3
Apr.	198.5	4.9	0.4	3.0	8.0	257.2	SSE	5.6	0.0
May	235.0	6.4	1.3	4.1	8.3	280.8	S	5.3	0.0
June	148.5	7.6	2.1	5.0	8.9	274.8	S	4.5	0.0
July	126.0	11.9	2.4	7.7	13.2	231.2	SSE	4.3	0.0
Aug.	133.5	10.5	2.0	1.3	9.4	238.6	SSE	3.7	0.0
Sept.	161.0	7.3	1.9	0.1	6.7	245.0	S	4.1	0.0
Oct.	123.5	6.0	1.1	0.3	4.3	241.7	S	4.7	0.0
Nov.	90.0	6.0	0.3	0.4	4.8	184.3	NNW	5.2	0.0
Dec.	56.5	5.2	0.0	0.7	4.4	173.3	NNW	4.9	0.0
Annual	1511.0	76.0	11.5	27.7	82.8	2763.5	S	4.8	0.3

[1] Mean daily temperature: 1905–1940, 1950–1952; mean precipitation: 1905–1940, 1950–1952.

TABLE XXXVII

CLIMATIC TABLE FOR PAOTING
Latitude 38°53′N, longitude 115°34′E, elevation 21.9 m

Month	Mean sta. press. (mbar)	Mean daily temp.[1] (°C)	Mean daily temp. range (°C)	Temp. extremes		Relative humidity (%)	Mean precip.[1] (mm)	Max. precip. (24 h)	Max. snow cover (cm)
				highest (°C)	lowest (°C)				
Jan.		− 5.5	10.9	14.9	−24.5	63	3.7	5.1	7.1
Feb.		− 1.3	12.9	18.0	−20.0	60	2.8	15.0	4.8
Mar.		5.7	13.3	26.0	−13.9	51	6.8	10.0	2.0
Apr.		14.1	14.0	35.1	− 2.2	48	8.6	18.0	0.0
May		20.4	15.0	43.0	5.0	49	20.0	72.7	0.0
June	no	25.7	14.1	42.1	9.0	53	69.5	71.0	0.0
July	data	26.9	11.0	41.9	14.8	72	139.5	110.0	0.0
Aug.		25.3	10.8	36.7	14.0	79	149.6	167.6	0.0
Sept.		20.2	13.0	34.9	2.5	67	37.6	56.0	0.0
Oct.		13.2	15.0	31.0	− 3.0	58	8.1	40.5	0.0
Nov.		4.2	12.5	25.0	−10.5	58	10.0	14.9	1.5
Dec.		− 2.8	11.6	12.4	−17.8	62	5.2	11.2	2.3
Annual		12.1	12.9	43.0	−24.5	60	443.7	167.9	7.1

Month	Mean evap. (mm)	Number of days			Number of cloudy days	Mean sunshine (h)	Most freq. wind direction	Mean wind speed (m/sec)	Number of days sand-storm
		precip.	thunder-storm	fog					
Jan.	44.5	2.6	0.0	4.0	5.4	178.2	N	1.1	1.0
Feb.	89.0	1.5	0.0	3.2	7.0	175.8	SW	1.3	0.7
Mar.	135.5	2.7	0.1	2.4	8.0	230.4	SSW	2.2	2.7
Apr.	298.0	2.8	0.3	0.4	9.2	227.9	SSW	2.1	5.7
May	330.5	5.1	2.6	1.6	12.2	271.2	SW	2.0	1.3
June	345.0	6.7	4.4	0.4	10.4	262.8	SSW	1.7	2.7
July	238.5	12.0	5.0	1.6	15.8	226.8	SSW	1.1	2.0
Aug.	170.5	10.7	2.9	1.2	14.4	206.4	SW, SSW	0.7	0.7
Sept.	169.0	5.1	0.8	5.6	9.6	223.9	SW	0.8	0.0
Oct.	80.0	2.3	0.4	4.4	6.0	233.4	SSW	1.0	1.0
Nov.	56.0	2.0	0.0	10.4	8.6	170.6	NNE	1.2	0.0
Dec.	30.0	2.5	0.0	5.6	5.8	165.9	SSW	0.9	0.7
Annual	1986.5	53.9	16.5	40.8	112.4	2569.4	SSW	1.3	18.5

[1] Mean daily temperature: 1913–1937, 1952; mean precipitation: 1914–1937, 1944–1950.

TABLE XXXVIII

CLIMATIC TABLE FOR TAIYUAN
Latitude 37°55′N, longitude 112°34′E, elevation 781.6 m

Month	Mean sta. press. (mbar)	Mean daily temp.[1] (°C)	Mean daily temp. range (°C)	Temp. extremes highest (°C)	Temp. extremes lowest (°C)	Relative humidity (%)	Mean precip.[1] (mm)	Max.. precip. (24 h)	Max. snow cover (cm)
Jan.	931.1	− 7.3	15.0	14.7	−29.7	56	4.0	30.0	9.0
Feb.	930.1	− 3.2	14.3	19.4	−22.3	59	3.8	8.3	13.2
Mar.	927.5	4.2	14.7	28.0	−18.3	54	11.0	74.8	5.5
Apr.	925.8	11.6	15.7	33.5	−15.0	55	14.8	28.9	0.0
May	923.3	18.5	15.8	38.2	0.0	53	26.6	51.3	0.0
June	931.5	22.8	14.9	39.8	4.6	56	39.5	70.8	0.0
July	919.7	25.0	12.3	41.4	8.2	67	107.6	66.2	0.0
Aug.	923.5	22.8	11.9	37.2	6.9	71	104.1	72.5	0.0
Sept.	927.9	17.4	14.6	34.6	− 0.9	67	49.8	49.8	0.0
Oct.	931.8	10.4	16.0	32.6	−12.0	61	18.4	36.2	0.0
Nov.	931.1	2.2	13.9	27.0	−20.1	64	10.9	35.2	3.3
Dec.	931.7	− 4.8	14.1	14.0	−24.0	63	3.9	10.4	15.0
Annual	927.0	10.0	14.5	41.4	−29.7	61	395.0	74.8	15.0

Month	Mean evap. (mm)	Number of days precip.	Number of days thunder-storm	Number of days fog	Number of cloudy days	Mean sunshine (h)	Most freq. wind direction	Mean wind speed (m/sec)	Number of days sand-storm
Jan.	52.0	1.2	0.0	2.1	2.7	170.5	SSE	1.2	2.0
Feb.	83.0	2.3	0.0	0.9	7.8	163.0	NW	1.5	1.3
Mar.	110.0	2.6	0.0	0.4	8.7	215.5	WNW	1.9	1.7
Apr.	246.0	3.3	1.2	0.2	8.3	205.9	WNW	2.2	5.3
May	304.5	5.0	4.6	0.3	11.0	235.8	N	1.9	2.3
June	312.0	6.5	8.4	0.2	5.0	234.3	N	1.7	1.0
July	239.0	11.0	10.0	0.5	14.0	203.8	SSE	1.4	0.7
Aug.	186.0	9.3	7.8	1.5	11.0	201.2	N	1.0	0.0
Sept.	158.0	7.2	3.6	2.1	7.3	209.5	NNE	1.3	0.3
Oct.	87.0	3.0	0.2	2.2	6.3	220.0	SSE	1.2	0.0
Nov.	51.0	2.4	0.0	3.1	10.3	166.4	WNW	1.2	0.0
Dec.	43.5	1.8	0.0	3.5	4.0	155.9	WNW	1.1	0.7
Annual	1872.0	55.8	35.8	16.8	95.7	2382.3	SSE	1.5	15.3

[1] Mean daily temperature: 1924–1938, 1950–1952; mean precipitation: 1916–1952.

TABLE XXXIX

CLIMATIC TABLE FOR SEOUL[1]

Latitude 37°34'N, longitude 126°58'E, elevation 85.5 m

Month	M.S.L. press. (mbar)	Mean daily temp. (°C)	Mean daily temp. range (°C)	Numb. of days with		Relative humidity (%)	Mean precip. (mm)	Number of days with snow cover	Max. snow cover (cm)
				max. temp. (<0°C)	min. temp. (<0°C)				
Jan.	1025.4	− 4.9	9.2	16	30	64	17.1	12	27.5
Feb.	1023.4	− 1.9	9.3	8	26	64	21.0	7	24.4
Mar.	1020.3	3.6	9.8	1	18	64	55.6	3	12.2
Apr.	1015.6	10.5	11.3	–	2	63	68.1	–	2.3
May	1011.8	16.3	11.6	–	–	66	86.3	–	–
June	1007.9	20.8	10.2	–	–	73	169.3	–	–
July	1007.1	24.5	8.2	–	–	81	358.0	–	–
Aug.	1008.7	25.4	8.8	–	–	78	224.2	–	–
Sept.	1013.2	20.3	10.4	–	–	73	142.3	–	–
Oct.	1020.1	13.4	11.9	–	1	68	49.2	–	–
Nov.	1023.3	6.3	10.4	1	11	68	36.0	1	8.5
Dec.	1025.0	− 1.2	8.6	8	27	66	32.0	5	6.6
Annual	1016.8	11.1	10.0	34	115	69	1259.2	28	27.5

Month	Soil temp. at 50 cm (°C)	Number of days with rain (≥ 1.0 mm)	Number of clear days	Number of days with fog	Number of cloudy days	Mean sunshine (h)	Number of days with rain (≥ 10 mm)	Mean max. temp. (°C)	Dew point temp. (°C)
Jan.	0.5	3	12	11.5	6	180.4	–	− 0.4	− 9.4
Feb.	0.1	3	11	9.2	6	181.7	–	2.8	− 7.1
Mar.	3.1	6	7	7.3	8	207.4	1	8.7	− 2.2
Apr.	9.5	6	7	5.5	9	227.5	2	16.5	3.6
May	15.8	7	5	5.1	11	256.8	3	22.5	9.5
June	20.6	9	3	6.3	14	213.9	4	26.5	15.2
July	24.4	14	2	10.1	19	179.3	8	29.2	20.7
Aug.	26.1	10	3	6.0	15	201.5	5	30.5	20.9
Sept.	22.8	7	6	6.4	11	205.6	3	26.1	15.1
Oct.	16.9	5	11	6.1	6	231.2	2	19.9	7.5
Nov.	10.2	5	10	12.2	6	179.6	1	11.7	1.5
Dec.	3.9	5	10	12.1	6	160.5	1	3.2	− 5.7
Annual	12.8	80	87	97.8	117	2425.9	30	16.4	8.9

[1] Period: 1931–1960.

TABLE XL

CLIMATIC TABLE FOR TSINAN
Latitude 36°41′N, longitude 116°58′E, elevation 55.1 m

Month	M.S.L. press. (mbar)	Mean daily temp.[1] (°C)	Mean daily temp. range (°C)	Temp. extremes		Relative humidity (%)	Mean precip.[1] (mm)	Max. precip. (24 h)	Max. snow cover (cm)
				highest (°C)	lowest (°C)				
Jan.	1025.3	− 1.2	9.7	20.2	−19.2	65	7.1	16.4	5.8
Feb.	1022.5	1.4	10.0	20.5	−15.2	66	8.9	15.8	12.8
Mar.	1017.8	8.4	11.4	32.3	−11.3	56	12.0	23.2	5.5
Apr.	1011.5	16.1	12.2	36.0	− 2.3	50	20.9	67.4	0.0
May	1005.8	22.5	12.6	40.8	2.5	54	34.3	95.3	0.0
June	1001.5	27.3	11.8	42.4	10.5	54	71.9	67.4	0.0
July	999.8	28.4	9.9	42.7	16.9	74	184.9	146.9	0.0
Aug.	1002.5	26.6	9.3	41.2	13.2	79	174.0	124.6	0.0
Sept.	1010.3	22.2	11.1	36.8	4.4	69	64.2	49.5	0.0
Oct.	1016.9	16.3	12.1	34.7	− 2.7	64	21.4	39.8	0.0
Nov.	1020.3	8.3	10.7	28.2	−11.3	66	19.9	36.0	2.5
Dec.	1024.2	1.0	9.6	19.0	−16.7	68	11.9	24.4	6.0
Annual	1013.1	14.8	10.9	42.7	−19.2	64	631.3	146.9	12.8

Month	Mean evap. (mm)	Number of days			Number of cloudy days	Mean sunshine (h)	Most freq. wind direction	Mean wind speed (m/sec)	Number of days sand-storm
		precip.	thunder-storm	fog					
Jan.	77.0	3.1	0.0	3.1	4.4	183.9	SW	2.5	0.0
Feb.	161.0	3.5	0.0	2.4	9.0	179.3	ENE	3.0	0.6
Mar.	192.0	3.9	0.2	1.6	6.2	220.0	SW	3.6	0.8
Apr.	399.5	4.6	1.0	0.9	8.0	232.4	SW	3.9	1.8
May	466.0	5.8	2.5	0.9	9.8	270.6	SW	3.6	1.0
June	432.0	7.7	4.7	1.0	7.0	275.4	NE	3.3	1.0
July	274.5	12.7	6.9	5.0	12.8	243.8	NE	2.5	0.2
Aug.	213.0	12.5	5.7	3.4	10.6	227.6	NE	1.9	0.0
Sept.	221.5	6.6	1.7	3.4	9.2	227.1	ENE	3.1	0.0
Oct.	151.0	3.2	0.5	1.9	8.0	241.4	SW	2.6	0.0
Nov.	98.5	3.7	0.0	3.3	7.2	187.9	ENE	2.9	0.0
Dec.	75.0	3.7	0.0	1.7	6.2	178.9	SW	2.5	0.0
Annual	2761.0	71.0	28.3	27.3	100.4	2168.4	SW	2.9	5.4

[1] Mean daily temperature: 1919–1952; mean precipitation: 1916–1952.

TABLE XLI

CLIMATIC TABLE FOR HSINING
Latitude 36°35′N, longitude 101°55′E, elevation 2244.2 m

Month	Mean sta. press. (mbar)	Mean daily temp.[1] (°C)	Mean daily temp. range (°C)	Temp. extremes		Relative humidity (%)	Mean precip.[1] (mm)	Max. precip. (24 h)	Max. snow cover (cm)
				highest (°C)	lowest (°C)				
Jan.		− 6.4	16.4	12.1	−23.1	50	1.1	3.4	2.0
Feb.		− 2.6	15.3	19.1	−19.4	45	1.8	5.4	2.0
Mar.		2.9	14.4	25.0	−13.4	40	5.3	6.2	4.5
Apr.		8.6	14.6	27.9	−11.2	45	18.4	25.0	2.0
May		13.3	15.0	30.9	− 2.3	48	33.5	21.0	1.0
June	no	15.7	13.3	32.1	2.7	58	45.9	27.1	0.0
July	data	18.3	12.8	32.4	6.1	61	72.7	34.0	0.0
Aug.		17.8	11.7	32.3	5.5	65	92.2	47.0	0.0
Sept.		12.9	10.5	27.1	0.1	68	73.8	39.1	0.0
Oct.		7.9	12.2	25.6	− 6.6	62	26.8	25.3	Tr.
Nov.		0.1	13.7	20.7	−18.9	56	4.1	12.2	2.0
Dec.		− 5.3	15.5	13.1	−22.6	54	1.5	2.6	2.7
Annual		6.9	13.8	32.4	−23.1	54	377.2	47.0	4.5

Month	Mean evap. (mm)	Number of days			Number of cloudy days	Mean sunshine (h)	Most freq. wind direction	Mean wind speed (m/sec)	Number of days sand-storm
		precip.	thunder-storm	fog					
Jan.		1.6	0.0	0.0	3.0	213.3	E	1.9	1.0
Feb.		1.8	0.0	0.0	5.9	200.9	E	1.5	0.3
Mar.		3.1	0.0	0.4	10.7	194.3	E	1.7	0.5
Apr.		4.0	0.1	0.5	10.1	214.3	E	1.9	1.3
May		8.0	2.7	0.5	10.4	245.7	E	1.4	1.0
June	no	11.6	3.8	0.1	11.1	242.6	E	1.3	0.0
July	data	12.3	5.2	0.1	12.6	239.1	E	1.2	0.3
Aug.		11.9	3.6	0.3	12.4	233.4	E	1.3	0.3
Sept.		11.9	2.8	0.3	12.6	188.0	E	1.2	0.0
Oct.		7.6	1.0	0.3	8.9	209.2	E	1.3	0.7
Nov.		1.9	0.0	0.1	4.0	211.1	E	1.2	1.0
Dec.		1.3	0.0	0.0	3.1	226.7	SE	0.9	0.3
Annual		76.9	19.3	2.5	104.8	2618.9	E	1.3	6.7

[1] Mean daily temperature: 1936–1949; mean precipitation: 1936–1948.

TABLE XLII

CLIMATIC TABLE FOR TSINGTAO
Latitude 36°04′N, longitude 120°19′E, elevation 77.0 m

Month	M.S.L. press. (mbar)	Mean daily temp.[1] (°C)	Mean daily temp. range (°C)	Temp.extremes highest (°C)	Temp.extremes lowest (°C)	Relative humidity (%)	Mean precip.[1] (mm)	Max. precip. (24 h)	Max. snow cover (cm)
Jan.	1028.6	− 1.1	7.6	11.6	−16.4	68	10.9	33.4	2.0
Feb.	1026.1	0.1	7.9	15.4	−12.8	68	9.2	23.2	3.0
Mar.	1021.9	4.4	7.3	22.6	−11.3	68	18.9	55.5	0.0
Apr.	1015.9	10.3	7.5	29.7	− 4.3	70	33.5	69.0	0.0
May	1011.0	15.7	7.5	31.5	3.2	74	41.5	67.2	0.0
June	1006.3	20.0	6.4	33.2	10.9	82	76.9	136.3	0.0
July	1005.4	23.7	5.2	36.2	14.5	89	149.6	132.3	0.0
Aug.	1006.7	25.1	4.0	35.6	14.6	84	149.3	162.5	0.0
Sept.	1013.9	21.4	7.1	33.7	8.7	73	84.7	225.4	0.0
Oct.	1020.5	15.9	7.6	28.5	1.1	66	33.1	132.6	0.0
Nov.	1024.3	8.6	7.5	28.8	− 9.2	65	22.2	74.0	0.0
Dec.	1027.7	1.6	6.6	15.8	−14.1	67	16.8	21.9	3.0
Annual	1017.4	12.1	6.9	36.2	−16.4	73	647.3	225.4	3.0

Month	Mean evap. (mm)	Number of days precip.	Number of days thunder-storm	Number of days fog	Number of cloudy days	Mean sunshine (h)	Most freq. wind direction	Mean wind speed (m/sec)	Number of days sand-storm
Jan.	51.5	3.8	0.0	2.4	5.8	182.3	N	5.4	0.5
Feb.	76.5	3.7	0.1	2.7	6.3	174.9	N	5.3	0.0
Mar.	100.5	4.5	0.3	3.7	8.0	211.0	S	5.8	0.5
Apr.	165.0	5.6	0.8	5.1	8.6	223.5	SSE	5.7	0.5
May	198.5	6.9	2.1	7.5	9.3	246.6	SSE	5.7	0.3
June	151.5	7.9	2.5	9.3	11.4	217.0	SSE	5.5	0.0
July	142.0	12.6	3.9	11.3	14.9	187.2	SSE	5.1	0.0
Aug.	166.0	11.2	3.8	2.1	10.7	219.6	SSE	4.6	0.0
Sept.	166.5	7.6	1.4	0.6	8.2	212.1	N	4.7	0.2
Oct.	153.5	4.3	0.6	0.7	5.1	229.2	N	5.0	0.0
Nov.	93.0	4.4	0.1	0.9	5.3	187.4	N	5.5	0.0
Dec.	63.5	4.4	0.0	1.7	6.1	170.6	N	5.6	0.0
Annual	1528.0	76.9	15.6	48.1	99.8	2462.0	S	5.3	2.0

[1] Mean daily temperature: 1898–1952; mean precipitation: 1898–1952.

TABLE XLIII

CLIMATIC TABLE FOR LANCHOW
Latitude 36°01′N, longitude 103°59′E, elevation 1507.8 m

Month	Mean sta. press. (mbar)	Mean daily temp.[1] (°C)	Mean daily temp. range (°C)	Temp. extremes		Relative humidity (%)	Mean precip.[1] (mm)	Max. precip. (24 h)	Max. snow cover (cm)
				highest (°C)	lowest (°C)				
Jan.		− 6.5	15.4	13.8	−23.1	61	1.3	16.0	4.0
Feb.		− 1.7	15.3	17.5	−22.1	54	3.1	9.0	4.5
Mar.		5.4	14.3	26.9	−16.3	47	7.9	8.6	4.0
Apr.		12.1	14.8	33.2	− 8.4	45	14.0	14.8	1.0
May		17.4	15.0	35.5	0.4	47	34.1	39.0	0.0
June		20.9	14.1	36.5	2.9	52	39.7	52.5	0.0
July		22.8	12.9	38.0	9.3	60	65.8	69.0	0.0
Aug.		21.4	12.5	37.0	5.4	64	92.2	71.8	0.0
Sept.	no	16.3	11.5	31.9	0.5	69	55.5	70.3	0.0
Oct.	data	10.1	12.6	28.0	− 6.6	67	18.4	19.9	1.3
Nov.		1.7	13.4	21.6	−15.4	63	3.8	13.6	1.0
Dec.		− 5.3	14.3	10.9	−21.6	64	1.8	9.1	6.0
Annual		9.5	13.8	38.0	−23.1	58	337.6	71.8	6.0

Month	Mean evap. (mm)	Number of days			Number of cloudy days	Mean sunshine (h)	Most freq. wind directions	Mean wind speed (m/sec)	Number of days sand-storm
		precip.	thunder-storm	fog					
Jan.	30.0	1.2	0.0	0.0	6.5	159.6	E	1.0	0.0
Feb.	46.5	2.2	0.1	0.5	8.2	143.3	ENE	1.2	0.5
Mar.	102.5	4.3	0.2	0.0	13.4	146.9	E	1.8	1.6
Apr.	170.0	5.3	1.1	0.3	13.2	169.6	E	1.9	2.4
May	243.0	7.4	2.4	0.3	12.7	201.6	E	1.9	2.9
June	263.0	8.8	4.1	0.0	14.0	208.9	E	1.9	1.5
July	257.0	10.4	4.4	0.0	13.2	224.3	E	1.8	1.1
Aug.	194.0	11.3	3.8	0.3	11.6	226.6	E	1.7	0.6
Sept.	137.5	11.7	2.6	0.3	13.6	159.4	SW	1.5	0.3
Oct.	65.0	6.1	0.3	1.5	12.4	166.5	E	1.3	0.2
Nov.	37.5	1.8	0.0	0.3	7.7	166.2	E	1.0	0.2
Dec.	19.0	1.2	0.0	0.5	5.0	167.9	E	1.0	0.1
Annual	1565.0	71.7	18.7	3.8	131.5	2146.6	E	1.5	11.4

[1] Mean daily temperature: 1932–1952; mean precipitation: 1932–1952.

TABLE XLIV

CLIMATIC TABLE FOR PUSAN[1]

Latitude 35°06′N, longitude 129°02′E, elevation 69.2 m

Month	M.S.L. press. (mbar)	Mean daily temp. (°C)	Mean daily temp. range (°C)	Numb. of days with max. temp. (<0°C)	min. temp. (<0°C)	Relative humidity (%)	Mean precip. (mm)	Number of days with snow cover	Max. snow cover (cm)
Jan.	1022.3	1.8	8.3	2	20	49	25.3	1	6.4
Feb.	1021.2	3.5	8.5	1	15	52	44.1	1	22.5
Mar.	1019.0	7.3	8.6	–	6	59	88.5	–	0.8
Apr.	1015.6	12.5	8.2	–	–	66	113.5	–	
May	1011.9	16.7	7.8	–	–	71	139.3	–	–
June	1008.3	19.8	6.6	–	–	80	197.5	–	–
July	1007.9	23.7	5.9	–	–	85	247.6	–	–
Aug.	1009.0	25.4	6.5	–	–	80	165.0	–	–
Sept.	1012.5	21.6	7.2	–	–	74	205.1	–	–
Oct.	1018.6	16.6	8.5	–	–	64	73.1	–	–
Nov.	1021.6	11.1	8.6	–	1	59	43.9	–	0.4
Dec.	1022.7	5.0	8.2	1	12	53	38.5	–	17.0
Annual	1015.8	13.8	7.8	4	55	66	1381.4	2	22.5

Month	Soil temp. at 50 cm (°C)	Number of days with rain (⩾ 1.0 mm)	Number of clear days	Number of days with fog	Number of cloudy days	Mean sunshine (h)	Number of days with rain (⩾ 10 mm)	Mean max. temp. (°C)	Dew point temp. (°C)
Jan.	5.0	3	13	1.0	5	205.1	1	6.4	− 6.4
Feb.	4.3	4	11	1.3	7	190.0	1	8.2	− 4.4
Mar.	8.4	6	9	3.4	10	212.6	3	12.1	0.6
Apr.	12.8	7	7	7.5	11	218.7	4	17.1	6.4
May	17.5	8	6	13.3	12	241.0	4	21.3	11.2
June	21.0	9	3	17.2	17	195.7	5	23.7	16.0
July	24.2	11	3	18.6	19	183.0	6	27.2	21.0
Aug.	26.7	8	5	3.2	13	231.6	4	29.4	21.5
Sept.	23.8	9	5	1.3	14	181.5	5	25.8	16.8
Oct.	19.3	4	10	1.1	8	218.0	2	21.5	9.8
Nov.	14.2	4	11	0.5	6	195.8	2	16.0	3.9
Dec.	8.4	4	14	0.5	5	198.4	1	9.5	− 2.5
Annual	15.6	77	102	68.9	127	2471.2	38	18.2	10.3

[1] Period: 1934–1960.

TABLE XLVII

CLIMATIC TABLE FOR NANKING
Latitude 32°04′N, longitude 118°47′E, elevation 61.5 m

Month	Mean sta. press. (mbar)	Mean daily temp.[1] (°C)	Mean daily temp. range (°C)	Temp. extremes		Relative humidity (%)	Mean precip.[1] (mm)	Max. precip. (24 h)	Max. snow cover (cm)
				highest (°C)	lowest (°C)				
Jan.	1019.7	2.2	7.8	19.6	−13.8	74	38.8	29.0	51.0
Feb.	1016.6	4.0	7.6	23.0	− 7.4	77	55.9	24.1	13.0
Mar.	1012.7	8.8	9.4	29.4	− 6.0	70	74.2	54.6	18.2
Apr.	1007.1	15.0	9.8	34.0	0.4	72	81.6	53.3	0.0
May	1002.5	20.5	9.6	35.7	6.8	74	92.8	78.5	0.0
June	997.4	24.7	9.0	33.2	15.0	76	118.1	125.1	0.0
July	995.8	28.0	8.1	43.0	16.8	80	152.2	198.5	0.0
Aug.	997.0	28.0	8.6	40.9	18.5	79	163.9	64.4	0.0
Sept.	1004.5	23.3	8.9	38.8	10.3	78	64.0	81.4	0.0
Oct.	1011.8	17.5	9.5	34.0	3.7	71	46.6	39.5	0.0
Nov.	1015.3	11.2	9.1	28.1	− 4.5	71	37.9	30.7	7.1
Dec.	1018.5	4.9	7.5	24.5	−12.6	76	52.2	40.6	50.0
Annual	1008.2	15.7	8.7	43.0	−13.8	75	918.3	198.5	51.0

Month	Mean evap. (mm)	Number of days			Number of cloudy days	Mean sunshine (h)	Most freq. wind direction	Mean wind speed (m/sec)	Number of days sand-storm
		precip.	thunder-storm	fog					
Jan.	46.0	8.5	0.1	4.5	13.2	137.0	NNE	4.5	0.0
Feb.	75.5	10.3	0.4	4.5	15.5	110.2	NE	4.9	0.0
Mar.	88.5	10.0	1.0	2.4	15.9	154.6	NE, ESE	5.4	0.0
Apr.	154.0	11.3	2.1	3.2	16.4	152.7	ESE	5.2	0.0
May	195.5	11.3	1.9	2.1	17.0	204.9	ESE	4.8	0.3
June	185.0	10.4	2.9	1.7	17.2	191.8	ESE	4.7	0.0
July	273.0	13.4	6.8	0.9	15.9	231.2	ESE	4.8	0.0
Aug.	230.5	12.3	6.3	1.1	14.1	231.8	ESE	4.5	0.0
Sept.	151.0	10.4	1.5	2.7	14.1	183.1	NE	4.0	0.0
Oct.	130.0	7.7	0.2	3.0	12.2	190.0	NNE, ESE	4.2	0.0
Nov.	67.5	8.6	0.1	3.7	13.1	142.6	NNE	4.5	0.0
Dec.	52.0	10.6	0.0	4.9	14.6	127.8	NNE	4.5	0.0
Annual	1648.5	124.8	23.4	34.7	179.4	2057.6	NE, ESE	4.7	0.3

[1] Mean daily temperature: 1929–1936, 1946–1953; mean precipitation: 1929–1953.

TABLE XLVIII

CLIMATIC TABLE FOR HOFEI
Latitude 31°53′N, longitude 117°15′E, elevation 25.7 m

Month	Mean sta. press. (mbar)	Mean daily temp.[1] (°C)	Mean daily temp. range (°C)	Temp. extremes		Relative humidity (%)	Mean precip.[1] (mm)	Max. precip. (24 h)	Max. snow cover (cm)
				highest (°C)	lowest (°C)				
Jan.		0.5	7.1	17.5	−20.6	80	34.3	28.6	39.0
Feb.		5.0	6.8	19.9	− 6.7	82	36.3	19.2	24.0
Mar.		8.5	7.6	28.1	− 1.4	77	118.9	47.1	0.0
Apr.		15.2	8.4	31.6	3.8	75	49.4	22.2	0.0
May		21.3	9.0	33.7	11.1	73	97.7	45.2	0.0
June	no	24.6	7.9	37.1	15.8	73	87.0	102.3	0.0
July	data	27.4	6.2	36.3	20.7	85	98.4	94.5	0.0
Aug.		28.3	8.1	37.9	17.2	80	92.2	129.3	0.0
Sept.		23.6	8.3	37.0	12.2	76	109.8	101.0	0.0
Oct.		17.1	9.5	29.5	4.5	71	37.1	22.2	0.0
Nov.		10.8	10.0	25.2	− 4.3	73	36.7	19.8	0.0
Dec.		4.3	8.4	20.2	− 5.2	77	32.2	18.8	45.0
Annual		15.5	8.1	37.9	−20.7	77	830.1	129.3	45.0

Month	Mean evap. (mm)	Number of days			Number of cloudy days	Mean sunshine (h)	Most freq. wind direction	Mean wind speed (m/sec)	Number of days sand-storm
		precip.	thunder-storm	fog					
Jan.		8.3	0.0	4.0	14.5	128.4	NE	2.3	0.0
Feb.		3.7	0.3	1.3	12.0	114.3	ENE	3.0	0.0
Mar.		11.3	1.0	1.3	13.5	158.1	ENE	3.3	0.0
Apr.		7.0	1.0	1.0	17.0	178.3	SE	3.2	0.0
May		7.3	3.0	1.0	19.5	218.2	ESE	3.7	0.0
June	no	7.7	2.3	1.0	18.0	211.9	S	3.6	0.0
July	data	9.3	12.0	1.0	24.0	204.4	S	3.3	0.0
Aug.		8.3	8.0	0.7	10.5	293.8	E	2.9	0.0
Sept.		8.7	2.7	1.7	8.5	211.1	NE	2.5	0.0
Oct.		5.0	0.3	1.3	9.0	203.6	NE. E	2.6	0.0
Nov.		4.3	0.0	1.3	6.5	180.4	SE	3.1	0.0
Dec.		7.7	0.0	4.0	10.5	154.1	NE	3.2	0.0
Annual		88.7	30.7	10.7	163.5	2261.6	S	3.1	0.0

[1] Mean daily temperature: 1953–1955; mean precipitation: 1934, 1942–1948.

TABLE XLIX

CLIMATIC TABLE FOR SHANGHAI
Latitude 31°12′N, longitude 121°26′E, elevation 4.6 m

Month	Mean sta. press. (mbar)	Mean daily temp.[1] (°C)	Mean daily temp. range (°C)	Temp. extremes		Relative humidity (%)	Mean precip.[1] (mm)	Max. precip. (24 h)	Max. snow cover (cm)
				highest (°C)	lowest (°C)				
Jan.	1026.2	3.4	8.2	23.3	−12.1	78	469.2	55.3	3.8
Feb.	1024.3	4.3	8.0	28.5	−10.3	79	1.7	38.6	5.8
Mar.	1020.6	8.2	9.3	32.0	− 5.8	79	84.7	65.4	0.0
Apr.	1015.1	13.7	10.1	34.8	− 1.3	79	90.5	57.0	0.0
May	1010.6	18.9	10.2	35.7	3.0	80	95.9	89.9	0.0
June	1005.8	23.1	8.9	39.3	10.5	84	177.4	161.4	0.0
July	1004.3	27.1	8.9	40.2	15.9	84	147.8	148.2	0.0
Aug.	1004.3	27.2	9.2	40.0	10.1	84	138.7	155.3	0.0
Sept.	1011.8	23.0	8.9	37.8	6.8	83	131.6	195.5	0.0
Oct.	1018.6	17.7	10.2	33.6	1.1	79	74.3	108.2	0.0
Nov.	1023.1	11.6	9.8	29.8	− 5.1	78	53.3	74.4	0.0
Dec.	1025.5	5.9	8.7	24.1	−10.2	77	38.0	45.1	0.0
Annual	1015.9	15.3	9.2	40.2	−12.1	80	1143.0	195.5	5.8

Month	Mean evap. (mm)	Number of days			Number of cloudy days	Mean sunshine (h)	Most freq. wind direction	Mean wind speed (m/sec)	Number of days sand-storm
		precip.	thunder-storm	fog					
Jan.	51.5	9.5	0.0	2.8	12.7	123.4	NW	4.6	0.0
Feb.	67.0	10.2	0.2	2.9	18.3	104.6	NE	4.6	0.0
Mar.	77.5	12.2	1.0	3.4	17.3	136.3	ESE	4.9	0.0
Apr.	128.5	12.5	1.7	3.3	17.7	142.9	SE	4.9	0.0
May	148.0	12.1	1.9	3.7	17.0	168.7	SE	4.6	0.0
June	149.0	13.9	3.4	3.0	17.7	141.7	ESE	4.4	0.0
July	265.0	11.1	5.6	1.3	18.7	212.8	SE	4.9	0.0
Aug.	211.0	10.7	5.1	1.4	13.0	231.8	ESE	4.7	0.0
Sept.	156.5	11.8	1.5	3.4	18.9	158.0	ENE	4.1	0.0
Oct.	128.5	8.8	0.4	4.3	12.7	173.6	ESE	3.9	0.0
Nov.	74.0	8.2	0.2	4.3	15.3	140.6	NW	4.2	0.0
Dec.	62.5	7.6	0.1	3.8	12.3	137.7	NW	4.5	0.0
Annual	1519.0	128.8	21.2	36.9	190.7	1871.7	SE	4.5	0.0

[1] Mean daily temperature: 1873–1953; mean precipitation: 1873–1953.

TABLE L

CLIMATIC TABLE FOR CHANGTU
Latitude 31°11′N, longitude 96°59′E, elevation 3200 m

Month	Mean sta. press. (mbar)	Mean daily temp.[1] (°C)	Mean daily temp. range (°C)	Temp. extremes highest (°C)	lowest (°C)	Relative humidity (%)	Mean precip.[1] (mm)	Max. precip. (24 h)	Max. snow cover (cm)
Jan.		− 2.5	20.0	16.4	−18.0	46	1.6	2.0	2.5
Feb.		1.4	19.3	21.7	−15.8	44	4.1	4.8	4.5
Mar.		4.7	18.6	23.4	−12.2	45	10.6	5.8	2.0
Apr.		8.2	16.9	25.8	− 7.6	50	22.6	11.9	8.0
May		12.3	16.9	28.0	− 3.1	52	79.2	24.6	0.0
June	no	14.5	14.8	31.1	1.3	65	87.7	25.9	0.0
July	data	15.7	14.6	33.3	1.2	71	135.1	37.4	0.0
Aug.		15.0	15.1	33.1	0.2	70	98.5	24.0	0.0
Sept.		13.2	14.3	29.4	1.3	69	80.6	22.6	0.0
Oct.		9.1	16.8	26.9	− 6.8	61	28.8	22.5	0.0
Nov.		3.1	20.2	21.6	−12.7	43	0.8	2.4	0.7
Dec.		1.4	19.0	19.5	−17.8	45	3.7	1.6	10.0
Annual		7.8	17.2	33.3	−18.0	53	553.2	37.4	10.0

Month	Mean evap. (mm)	Number of days precip.	thunder-storm	fog	Number of cloudy days	Mean sunshine (h)	Most freq. wind direction	Mean wind speed (m/sec)	Number of days sand-storm
Jan.		3.0	0.0	0.0	3.5	181.9	C, NW	1.5	2.5
Feb.		2.8	0.0	0.0	4.5	179.7	C, N	1.4	2.3
Mar.		5.8	0.3	0.0	11.5	186.2	C, N	1.3	2.3
Apr.		9.3	3.0	0.0	13.5	185.7	C, N	1.4	1.0
May		15.8	5.7	0.0	10.5	226.7	C, N	1.7	0.8
June	no	19.0	5.3	0.0	15.0	179.8	C, N	1.5	0.0
July	data	22.5	10.3	0.0	16.5	155.8	C, NW	1.5	0.0
Aug.		19.5	8.3	0.0	11.5	203.3	C, SE	1.1	0.3
Sept.		18.5	8.7	0.0	11.0	175.6	C, NW	1.3	0.5
Oct.		10.3	0.7	0.0	6.0	200.2	C, S	1.2	0.5
Nov.		1.3	0.0	0.0	2.5	215.7	C, NNW	1.3	1.3
Dec.		1.5	0.0	0.0	3.5	191.8	C, NW	1.2	1.3
Annual		129.3	42.3	0.0	109.5	2282.4	C, NNW	1.4	12.8

[1] Mean daily temperature: 1952–1955; mean precipitation: 1952–1955.

TABLE LI

CLIMATIC TABLE FOR CHENGTU
Latitude 30°40′N, longitude 104°04′E, elevation 497.9 m

Month	Mean sta. press. (mbar)	Mean daily temp.[1] (°C)	Mean daily temp. range (°C)	Temp. extremes highest (°C)	Temp. extremes lowest (°C)	Relative humidity (%)	Mean precip.[1] (mm)	Max. precip. (24 h)	Max. snow cover (cm)
Jan.	928.6	6.2	6.9	21.3	− 4.0	81	7.4	7.5	1.0
Feb.	926.2	8.4	7.2	26.1	− 4.0	80	14.7	16.0	trace
Mar.	923.3	12.7	8.5	30.0	− 0.4	77	25.3	27.9	0.0
Apr.	918.7	17.5	9.5	38.4	0.0	76	56.3	91.1	0.0
May	914.6	22.2	9.7	37.9	9.6	74	96.0	63.6	0.0
June	909.0	24.4	8.4	38.3	13.1	80	121.8	84.4	0.0
July	905.3	26.5	8.6	40.1	17.1	84	304.3	233.0	0.0
Aug.	907.0	25.9	8.6	39.0	15.7	84	303.0	188.0	0.0
Sept.	917.1	22.1	7.0	34.5	12.9	85	139.1	129.9	0.0
Oct.	924.9	17.5	6.6	32.6	7.0	86	53.4	24.2	0.0
Nov.	927.1	12.4	6.5	26.2	0.1	84	17.6	11.8	0.0
Dec.	928.6	7.7	6.4	20.8	− 3.7	83	8.1	8.2	2.0
Annual	919.3	17.0	7.8	40.1	− 4.0	81	1146.1	233.0	2.0

Month	Mean evap. (mm)	Number of days precip.	Number of days thunder-storm	Number of days fog	Number of cloudy days	Mean sunshine (h)	Most freq. wind direction	Mean wind speed (m/sec)	Number of days sand-storm
Jan.	44.0	5.8	0.0	13.8	20.5	61.8	NNE	1.2	0.0
Feb.	46.5	7.6	0.0	8.4	20.2	60.3	NE	1.4	0.0
Mar.	65.5	10.7	0.3	6.9	21.5	81.5	NE	1.5	0.0
Apr.	94.5	13.9	1.7	7.5	20.5	101.6	NE	1.6	0.0
May	145.0	14.5	4.2	7.8	19.8	133.8	NE	1.6	0.0
June	132.0	15.9	3.0	5.2	21.0	111.4	SW	1.6	0.0
July	153.5	16.5	9.5	7.4	16.8	101.1	NNE	1.4	0.0
Aug.	134.5	15.9	9.4	9.0	15.5	171.4	NE	1.5	0.0
Sept.	104.5	16.2	1.4	8.0	22.8	76.5	N	1.6	0.0
Oct.	54.0	15.5	0.2	9.2	23.7	71.8	NNE	1.3	0.0
Nov.	43.5	8.7	0.0	12.8	21.2	59.5	NNE	1.3	0.0
Dec.	33.5	5.4	0.0	15.5	21.2	56.6	NNE	1.4	0.0
Annual	1051.0	146.4	29.7	111.4	244.6	1152.2	NE	1.4	0.0

[1] Mean daily temperature: 1932–1953; mean precipitation: 1932–1953.

TABLE LII

CLIMATIC TABLE FOR WUHAN
Latitude 30°33′N, longitude 114°17′E, elevation 23.0 m

Month	Mean sta. press. (mbar)	Mean daily temp.[1] (°C)	Mean daily temp. range (°C)	Temp. extremes		Relative humidity (%)	Mean precip.[1] (mm)	Max. precip. (24 h)	Max. snow cover (cm)
				highest (°C)	lowest (°C)				
Jan.	1024.6	3.8	7.5	21.1	−13.0	76	24.8	45.4	32.0
Feb.	1021.5	5.4	7.2	28.0	− 7.2	78	53.7	42.5	8.8
Mar.	1017.9	10.4	8.2	32.6	− 5.0	76	92.3	105.8	Tr.
Apr.	1012.1	16.3	8.9	36.0	0.3	77	140.2	166.3	0.0
May	1007.7	22.1	8.8	36.6	9.5	75	165.8	205.5	0.0
June	1002.2	25.8	8.1	38.2	14.3	77	215.5	219.9	0.0
July	1000.7	28.9	7.8	41.1	18.4	77	173.1	213.5	0.0
Aug.	1002.5	28.7	8.2	41.3	18.1	75	109.5	113.0	0.0
Sept.	1010.1	23.9	8.3	38.0	11.6	74	69.4	63.9	0.0
Oct.	1017.0	18.4	8.5	34.4	5.2	73	73.9	64.6	0.0
Nov.	1021.4	12.0	8.2	29.7	− 5.0	75	45.5	63.7	0.0
Dec.	1024.2	0.1	7.8	25.0	− 7.5	74	29.3	66.5	30.0
Annual	1013.4	16.8	8.1	41.3	−13.0	76	1202.0	219.9	32.0

Month	Mean evap. (mm)	Number of days			Number of cloudy days	Mean sunshine (h)	Most freq. wind direction	Mean wind speed (m/sec)	Number of days sand-storm
		precip.	thunder-storm	fog					
Jan.	49.5	8.2	0.5	6.8	11.7	117.1	NNE	2.0	0.0
Feb.	74.5	9.0	1.2	2.8	18.0	96.7	N	2.0	0.0
Mar.	71.0	11.7	2.3	4.0	17.0	126.5	NNE	2.0	0.0
Apr.	115.5	11.3	3.5	4.5	15.0	143.7	NE	2.0	0.0
May	158.0	12.6	4.8	2.5	15.7	184.5	SE	1.8	0.3
June	207.0	11.3	3.8	0.3	11.7	187.4	SE	1.7	0.0
July	244.5	10.0	6.2	0.0	14.7	240.1	SE	1.9	0.0
Aug.	243.5	7.5	9.0	0.3	9.7	262.2	NE	1.9	0.3
Sept.	155.5	7.6	2.5	4.0	11.7	192.6	NE	2.1	0.0
Oct.	125.5	8.8	0.2	4.3	17.0	164.1	N	1.8	0.0
Nov.	60.0	7.5	0.2	3.5	16.7	141.8	N	1.9	0.0
Dec.	54.5	6.7	0.2	11.8	11.0	120.4	NNE	1.9	0.0
Annual	1559.0	114.0	34.3	44.5	160.7	1967.0	NNE	1.9	0.6

[1] Mean daily temperature: 1905–1940, 1951–1953; mean precipitation: 1889–1940, 1951–1957.

TABLE LIII

CLIMATIC TABLE FOR HANGCHOW
Latitude 30°20′N, longitude 120°10′E, elevation 5.3 m

Month	M.S.L. press. (mbar)	Mean daily temp.[1] (°C)	Mean daily temp. range (°C)	Temp. extremes		Relative humidity (%)	Mean precip.[1] (mm)	Max. precip. (24 h)	Max. snow cover (cm)
				highest (°C)	lowest (°C)				
Jan.	1028.6	4.3	8.5	24.4	−10.5	81	67.5	44.8	14.0
Feb.	1024.2	5.2	7.6	27.2	− 7.5	82	95.1	51.6	11.2
Mar.	1021.4	9.4	8.8	31.6	− 4.0	81	123.8	88.1	0.0
Apr.	1015.3	15.3	9.1	37.0	1.2	80	136.7	60.9	0.0
May	1011.5	20.4	9.8	37.0	6.7	78	136.1	71.3	0.0
June	1005.9	24.7	8.3	39.7	10.1	82	236.3	129.1	0.0
July	1007.5	28.3	9.0	42.0	18.5	79	150.9	115.1	0.0
Aug.	1006.2	28.1	9.4	42.1	16.8	79	175.2	101.2	0.0
Sept.	1012.3	23.8	8.5	37.4	10.8	82	167.7	83.2	0.0
Oct.	1019.1	17.9	9.6	33.6	1.0	80	77.7	67.7	0.0
Nov.	1024.9	12.1	9.7	31.2	− 3.8	81	68.9	54.0	0.0
Dec.	1027.3	6.4	8.6	23.4	−10.5	80	59.8	46.8	2.0
Annual	1016.3	16.3	8.9	42.1	−10.5	80	1489.7	129.1	14.0

Month	Mean evap. (mm)	Number of days			Number of cloudy days	Mean sunshine (h)	Most freq. wind direction	Mean wind speed (m/sec)	Number of days sand-storm
		precip.	thunder-storm	fog					
Jan.	45.5	11.5	0.0	5.3	14.6	123.1	N	1.7	0.0
Feb.	56.5	13.2	0.3	3.7	16.5	96.5	N	1.0	0.0
Mar.	70.0	14.7	1.8	3.5	18.6	124.6	E	1.9	0.0
Apr.	113.5	14.5	2.3	4.0	18.1	120.0	E	1.8	0.0
May	138.0	13.1	1.6	3.8	16.7	179.7	E	1.8	0.0
June	126.5	16.0	3.1	2.7	17.8	135.9	E	1.5	0.0
July	263.0	12.0	6.6	1.7	12.8	221.3	E	1.6	0.0
Aug.	184.5	13.8	7.3	1.5	9.1	226.5	E	1.6	0.0
Sept.	123.5	13.4	2.0	3.7	15.8	146.3	NE	1.7	0.0
Oct.	109.5	10.2	0.2	6.2	15.0	151.8	NW	1.5	0.0
Nov.	65.0	9.7	0.1	7.7	12.8	135.2	NW	1.5	0.0
Dec.	51.5	10.4	0.0	5.3	13.6	121.8	NW	1.7	0.0
Annual	1347.0	152.6	26.6	49.0	181.4	1782.7	E	1.7	0.0

[1] Mean daily temperature: 1904–1947, 1950–1952; mean precipitation: 1904–1952.

TABLE LIV

CLIMATIC TABLE FOR CHUNGKING
Latitude 29°30′N, longitude 106°33′E, elevation 260.6 m

Month	Mean sta. press. (mbar)	Mean daily temp.[1] (°C)	Mean daily temp. range (°C)	Temp. extremes highest (°C)	Temp. extremes lowest (°C)	Relative humidity (%)	Mean precip.[1] (mm)	Max. precip. (24 h)	Max. snow cover (cm)
Jan.		8.1	5.1	20.6	− 0.8	83	17.4	29.2	0.0
Feb.		9.7	6.0	27.8	− 2.5	83	20.8	22.9	0.0
Mar.		14.1	7.3	31.7	3.1	81	38.1	40.9	0.0
Apr.		18.8	8.0	39.9	0.9	81	97.0	89.4	0.0
May		22.7	8.1	33.0	11.6	82	145.6	207.5	0.0
June	no	25.1	7.8	40.3	15.0	83	181.5	132.8	0.0
July	data	28.7	9.6	41.3	16.6	80	142.4	153.7	0.0
Aug.		28.8	7.1	44.0	17.8	77	120.0	97.5	0.0
Sept.		24.2	7.9	40.4	13.3	82	147.5	198.1	0.0
Oct.		18.7	6.0	34.6	6.8	87	109.1	70.0	0.0
Nov.		14.2	5.3	30.0	0.7	87	48.5	92.7	0.0
Dec.		9.9	5.0	21.7	− 1.5	86	20.7	18.0	0.0
Annual		18.6	7.2	44.0	− 2.5	83	1088.7	207.5	0.0

Month	Mean evap. (mm)	Number of days precip.	Number of days thunder-storm	Number of days fog	Number of cloudy days	Mean sunshine (h)	Most freq. wind direction	Mean wind speed (m/sec)	Number of days sand-storm
Jan.	30.5	6.7	0.2	14.6	20.2	47.9	N	1.1	0.0
Feb.	36.5	7.5	0.8	11.4	18.5	55.8	N	1.2	0.0
Mar.	75.5	10.1	1.0	10.4	20.3	80.9	N	1.3	0.0
Apr.	99.0	12.3	3.7	11.8	18.5	110.7	N	1.3	0.0
May	126.5	14.2	4.0	11.2	16.3	138.5	N	1.3	0.0
June	170.5	14.6	3.3	8.8	19.2	127.6	N	1.1	0.0
July	208.0	9.6	9.7	12.4	13.1	220.6	N	1.1	0.0
Aug.	217.5	9.2	6.5	8.5	10.3	221.9	N	1.0	0.0
Sept.	158.5	13.2	2.1	9.5	17.1	122.5	N	1.3	0.0
Oct.	50.0	16.2	0.5	11.3	21.9	62.9	N	0.9	0.0
Nov.	33.5	10.9	0.5	10.9	22.0	46.8	N	0.9	0.0
Dec.	22.5	7.6	0.0	13.5	22.4	53.8	N	0.8	0.0
Annual	1228.5	132.2	32.4	134.2	219.6	1280.8	N	1.1	0.0

[1] Mean daily temperature: 1924–1938, 1940–1953; mean precipitation: 1891–1953.

TABLE LV

CLIMATIC TABLE FOR NANCHANG
Latitude 28°40′N, longitude 115°58′E, elevation 48.9 m

Month	Mean sta. press. (mbar)	Mean daily temp.[1] (°C)	Mean daily temp. range (°C)	Temp. extremes		Relative humidity (%)	Mean precip.[1] (mm)	Max. precip. (24 h)	Max. snow cover (cm)
				highest (°C)	lowest (°C)				
Jan.		5.5	9.8	24.4	− 4.5	81	55.2	55.0	2.0
Feb.		7.0	11.2	27.2	− 5.9	85	108.1	59.5	5.0
Mar.		10.5	7.4	29.0	− 0.4	88	191.5	61.7	0.0
Apr.		16.4	7.3	33.8	6.1	85	250.2	131.7	0.0
May		21.4	8.2	36.5	10.6	84	288.6	118.9	0.0
June	no	25.1	7.6	39.0	13.9	84	295.1	151.9	0.0
July	data	28.9	8.4	39.0	20.0	82	257.5	88.5	0.0
Aug.		29.5	8.3	39.4	20.0	79	111.2	102.3	0.0
Sept.		24.6	9.9	38.7	14.8	82	108.9	63.5	0.0
Oct.		18.9	10.4	37.5	7.0	80	57.0	39.3	0.0
Nov.		13.6	9.6	29.7	0.4	83	70.4	58.0	0.0
Dec.		7.2	7.1	23.9	− 4.4	82	70.0	49.1	0.0
Annual		17.4	8.7	39.4	− 5.9	83	1769.9	151.9	5.0

Month	Mean evap. (mm)	Number of days			Number of cloudy days	Mean sunshine (h)	Most freq. wind direction	Mean wind speed (m/sec)	Number of days sand-storm
		precip.	thunder-storm	fog					
Jan.	62.5	8.7	0.0	5.2	15.3	121.0	N	3.8	0.0
Feb.	80.5	11.5	1.6	3.0	20.3	77.6	N	4.9	0.0
Mar.	79.5	14.3	3.8	3.4	22.3	92.4	N	3.5	0.3
Apr.	117.5	16.2	6.2	1.2	18.7	88.6	N	2.2	0.0
May	154.0	16.3	5.8	1.8	19.0	109.0	N	1.8	0.0
June	204.5	14.3	3.4	1.4	14.3	126.6	N	1.9	0.0
July	347.0	10.2	8.2	0.2	13.0	195.1	SSW	2.2	0.0
Aug.	305.0	9.0	8.6	0.4	6.7	275.8	N	2.2	0.7
Sept.	220.0	9.5	3.4	0.6	14.0	278.9	N	2.9	0.0
Oct.	236.5	7.6	0.2	2.6	20.3	234.4	N	2.8	0.0
Nov.	102.5	9.2	0.4	4.2	16.0	184.6	N	3.3	0.0
Dec.	83.5	9.9	0.6	3.6	14.0	159.1	N	3.2	0.0
Annual	1993.0	136.6	42.2	27.6	136.9	1939.2	N	2.9	1.0

[1] Mean daily temperature: 1929–1933, 1946–1953; mean precipitation: 1929–1953.

TABLE LVI

CLIMATIC TABLE FOR CHANGSHA
Latitude 28°15′N, longitude 112°50′E, elevation 48.0 m

Month	M.S.L. press. (mbar)	Mean daily temp.[1] (°C)	Mean daily temp. range (°C)	Temp. extremes		Relative humidity (%)	Mean precip.[1] (mm)	Max. precip. (24 h)	Max. snow cover (cm)
				highest (°C)	lowest (°C)				
Jan.	1025.3	4.3	5.9	26.7	− 8.1	84	64.3	38.1	3.0
Feb.	1022.2	6.2	5.8	26.2	− 3.0	87	120.9	55.0	2.0
Mar.	1018.3	10.9	7.4	28.9	− 1.2	85	122.4	56.6	0.0
Apr.	1012.7	16.8	7.4	35.3	4.3	84	202.8	92.0	0.0
May	1008.5	22.0	7.8	36.9	11.2	83	211.5	120.6	0.0
June	1004.1	25.6	7.8	38.9	15.3	81	253.7	195.0	0.0
July	1002.5	29.3	9.1	40.0	20.1	78	117.8	129.0	0.0
Aug.	1003.1	29.0	9.7	43.0	19.4	76	120.5	141.0	0.0
Sept.	1011.3	24.6	8.8	39.2	12.8	79	79.7	67.1	0.0
Oct.	1017.3	18.3	8.7	38.6	3.0	79	83.2	76.8	0.0
Nov.	1021.7	12.1	7.8	31.5	− 1.5	82	85.0	48.0	0.0
Dec.	1024.6	6.7	6.6	28.3	− 6.3	81	67.5	46.0	0.0
Annual	1014.3	17.2	7.7	43.0	− 8.1	82	1529.3	195.0	3.0

Month	Mean evap. (mm)	Number of days			Number of cloudy days	Mean sunshine (h)	Most freq. wind direction	Mean wind speed (m/sec)	Number of days sand-storm
		precip.	thunder-storm	fog					
Jan.	52.0	11.8	0.4	4.9	19.7	86.1	NNW	2.6	0.0
Feb.	69.5	14.5	1.4	3.8	20.7	51.1	NNW	3.1	0.0
Mar.	70.5	16.4	5.0	4.3	20.7	94.0	NNW	2.5	0.0
Apr.	110.5	17.5	7.6	4.5	19.6	73.1	NW	2.3	0.0
May	152.5	17.1	6.0	4.2	18.5	132.1	NW	2.1	0.0
June	228.5	14.1	3.4	2.5	16.6	144.1	S	2.0	0.3
July	364.5	10.1	8.6	1.5	12.6	205.7	S	2.4	0.0
Aug.	242.5	10.6	10.6	0.9	11.7	228.1	NW, S	2.1	0.3
Sept.	164.0	9.4	4.0	2.8	14.2	156.5	NW	2.5	0.0
Oct.	134.0	10.9	0.2	4.1	18.1	199.8	NW	2.3	0.0
Nov.	68.0	11.9	0.0	5.8	17.6	111.9	NNW, NW	2.6	0.0
Dec.	55.5	13.5	0.2	6.1	17.9	77.0	NNW	2.7	0.0
Annual	1712.0	157.8	47.4	45.4	208.2	1559.4	NW, NNW	2.5	0.7

[1] Mean daily temperature: 1932–1938, 1946–1950; mean precipitation: 1932–1950.

TABLE LVII

CLIMATIC TABLE FOR WENCHOW
Latitude 28°01′N, longitude 120°49′E, elevation 4.8 m

Month	M.S.L. press.[1] (mbar)	Mean daily temp.[1] (°C)	Mean daily temp. range (°C)	Temp. extremes		Relative humidity (%)	Mean precip.[1] (mm)	Max. precip. (24 h)	Max. snow cover (cm)
				highest (°C)	lowest (°C)				
Jan.	1026.7	7.7	5.5	23.0	− 3.0	80	49.3	48.3	1.0
Feb.	1024.2	8.4	5.7	23.0	− 2.2	85	89.1	70.6	7.8
Mar.	1021.9	11.6	6.6	27.2	− 0.5	86	132.4	76.2	0.0
Apr.	1016.6	16.7	7.1	32.5	5.5	86	147.7	72.6	0.0
May	1011.8	21.4	6.6	33.9	11.1	86	190.2	101.6	0.0
June	1007.0	25.2	6.1	38.3	15.5	89	264.5	107.2	0.0
July	1005.9	28.9	6.9	40.5	16.1	87	204.4	213.4	0.0
Aug.	1005.5	29.0	7.2	39.0	16.1	85	258.3	242.8	0.0
Sept.	1011.9	25.7	6.8	37.8	15.0	85	203.7	269.2	0.0
Oct.	1019.1	20.9	7.3	33.5	8.0	78	87.4	287.2	0.0
Nov.	1023.3	15.9	6.6	27.5	0.6	78	54.7	90.4	0.0
Dec.	1025.8	11.1	5.9	22.8	− 1.7	76	42.9	92.0	0.0
Annual	1016.6	18.5	6.6	40.5	− 3.0	83	1724.6	287.2	7.8

Month	Mean evap. (mm)	Number of days			Number of cloudy days	Mean sunshine (h)	Most freq. wind direction	Mean wind speed (m/sec)	Number of days sand-storm
		precip.	thunder-storm	fog					
Jan.	69.0	9.5	0.3	2.4	17.1	118.2	NW	3.4	0.0
Feb.	77.0	12.4	0.8	3.6	17.4	91.8	NW	3.2	0.0
Mar.	66.5	15.1	2.8	3.8	18.8	87.1	NW	3.0	0.0
Apr.	97.0	15.3	4.8	2.9	19.8	133.3	SE	2.8	0.0
May	124.5	16.1	4.3	1.9	19.9	103.8	SE	3.0	0.0
June	156.0	15.9	5.0	1.7	21.9	130.3	SE	2.7	0.0
July	243.0	12.9	10.5	0.6	11.6	212.5	SE	3.1	0.0
Aug.	197.5	13.4	12.0	0.1	8.8	241.7	SE	3.6	0.0
Sept.	150.5	12.2	6.3	0.5	14.5	200.6	NW	2.6	0.0
Oct.	164.0	8.6	0.0	0.7	13.8	162.8	NW	2.8	0.0
Nov.	96.0	7.2	0.5	1.0	15.6	145.4	NW	3.2	0.0
Dec.	88.5	7.5	0.3	1.3	16.6	134.0	NW	3.8	0.0
Annual	1529.5	145.8	47.3	20.5	196.0	1761.5	NW	3.1	0.0

[1] Mean daily temperature: 1924–1941; mean precipitation: 1883–1938, 1946–1950.

TABLE LVIII

CLIMATIC TABLE FOR KWEIYANG
Latitude 26°34′N, longitude 106°42′E, elevation 1071.2 m.

Month	Mean sta. press. (mbar)	Mean daily temp.[1] (°C)	Mean daily temp. range (°C)	Temp. extremes		Relative humidity (%)	Mean precip.[1] (mm)	Max. precip. (24 h)	Max. snow cover (cm)
				highest (°C)	lowest (°C)				
Jan.	900.4	5.0	7.5	26.0	− 9.5	81	21.2	21.8	Tr.
Feb.	898.3	6.6	8.1	30.0	− 5.4	80	27.5	25.9	0.0
Mar.	896.8	11.7	9.5	31.8	− 2.7	77	40.4	42.0	0.0
Apr.	894.2	16.4	10.1	35.1	− 2.3	77	93.9	88.2	0.0
May	892.3	20.4	9.7	35.1	6.7	78	200.1	125.3	0.0
June	890.1	22.5	8.8	35.6	10.1	79	204.1	94.6	0.0
July	889.0	24.7	9.2	39.5	13.5	76	187.2	79.0	0.0
Aug.	890.6	24.1	9.9	37.5	10.3	79	138.2	95.4	0.0
Sept.	895.7	21.0	9.8	36.4	5.4	78	120.1	82.5	0.0
Oct.	900.3	15.6	8.9	32.1	0.2	80	109.8	72.6	0.0
Nov.	901.3	11.7	8.1	29.7	− 5.7	81	47.4	34.3	0.0
Dec.	900.6	7.2	8.1	26.1	− 3.2	81	24.0	29.4	2.3
Annual	895.8	15.6	9.0	39.5	− 9.5	79	1214.0	125.3	2.3

Month	Mean evap. (mm)	Number of days			Number of cloudy days	Mean sunshine (h)	Most freq. wind direction	Mean wind speed (m/sec)	Number of days sandstorm
		precip.	thunderstorm	fog					
Jan.	48.0	13.8	0.3	8.2	23.1	58.1	N	1.7	0.0
Feb.	45.0	13.9	1.3	5.8	20.9	54.9	NE	1.9	0.0
Mar.	101.0	13.3	3.2	6.0	19.7	92.1	NE	2.1	0.0
Apr.	134.0	16.1	5.2	7.2	19.1	111.5	N	2.0	0.0
May	132.5	19.4	7.8	7.6	18.7	141.3	N	1.8	0.0
June	204.5	17.8	5.1	6.6	19.4	126.5	S	1.8	0.0
July	175.5	17.4	7.2	8.3	16.1	185.5	S	1.8	0.0
Aug.	173.5	15.8	7.9	10.6	13.7	185.5	S	1.6	0.0
Sept.	141.5	13.2	1.9	9.2	15.1	137.8	N	1.8	0.0
Oct.	81.0	15.8	0.5	9.0	20.2	84.6	N	1.7	0.0
Nov.	52.5	13.9	0.6	7.4	20.9	77.6	NE	1.9	0.0
Dec.	47.0	11.9	0.2	8.8	22.1	68.6	NE	1.8	0.0
Annual	1336.0	182.3	41.2	94.8	229.1	1323.8	N	1.8	0.0

[1] Mean daily temperature: 1920–1953; mean precipitation: 1921–1953.

TABLE LIX

CLIMATIC TABLE FOR FOOCHOW
Latitude 26°05′N, longitude 119°18′E, elevation 88.4 m

Month	M.S.L. press. (mbar)	Mean daily temp.[1] (°C)	Mean daily temp. range (°C)	Temp. extremes		Relative humidity (%)	Mean precip.[1] (mm)	Max. precip. (24 h)	Max. snow cover (cm)
				highest (°C)	lowest (°C)				
Jan.	1024.1	10.9	6.4	28.3	− 2.5	77	45.0	50.2	0.0
Feb.	1021.9	10.7	6.5	29.4	− 1.1	81	89.3	82.6	0.0
Mar.	1019.9	13.2	7.1	31.7	1.3	82	124.4	60.2	0.0
Apr.	1015.0	17.8	7.7	33.8	3.4	81	129.8	69.3	0.0
May	1010.7	22.6	7.1	35.8	10.6	84	100.7	95.3	0.0
June	1006.6	25.8	6.9	37.4	13.3	83	214.6	124.5	0.0
July	1005.3	28.5	7.9	39.8	20.2	82	170.3	239.6	0.0
Aug.	1004.9	28.5	7.9	38.4	20.0	81	188.5	201.9	0.0
Sept.	1010.2	26.1	7.3	37.8	14.6	81	193.0	283.3	0.0
Oct.	1016.9	22.1	7.7	35.2	4.2	77	49.0	101.6	0.0
Nov.	1020.9	18.0	7.3	33.4	3.9	77	41.4	81.3	0.0
Dec.	1022.9	13.7	7.2	28.7	0.4	77	44.4	61.0	0.0
Annual	1014.9	19.8	7.2	39.8	− 2.5	81	1450.4	283.3	0.0

Month	Mean evap. (mm)	Number of days			Number of cloudy days	Mean sunshine (h)	Most freq. wind direction	Mean wind speed (m/sec)	Number of days sand-storm
		precip.	thunder-storm	fog					
Jan.	83.5	7.5	0.2	3.6	18.6	112.8	NW	1.7	0.0
Feb.	88.0	11.4	0.6	6.8	18.0	103.1	SSE	1.7	0.0
Mar.	90.0	12.9	2.8	6.0	23.4	100.5	SSE	1.7	0.0
Apr.	112.0	12.4	6.6	7.6	20.3	126.8	SSE	1.6	0.0
May	127.0	13.1	7.8	5.8	21.4	127.8	SSE	1.6	0.0
June	191.5	13.1	10.2	6.0	21.4	156.2	SSE	1.6	0.0
July	236.0	9.5	14.0	1.6	17.9	241.9	SSE	1.9	0.0
Aug.	179.5	10.7	13.0	1.0	13.0	234.3	SSE	1.8	0.0
Sept.	154.5	11.3	7.2	2.4	17.6	191.9	E	1.8	0.0
Oct.	191.5	6.1	1.0	1.2	16.8	175.1	NNW	2.0	0.0
Nov.	110.0	5.8	0.0	2.6	7.6	146.4	NNW	1.9	0.0
Dec.	87.0	6.2	0.0	2.0	14.0	140.9	NNW	1.9	0.0
Annual	1650.5	120.1	63.4	46.6	220.0	1857.6	SSE	1.8	0.0

[1] Mean daily temperature: 1924–1944, 1951–1952; mean precipitation: 1880–1952.

TABLE LX

CLIMATIC TABLE FOR TENGCHUNG
Latitude 26°00′N, longitude 98°40′E, elevation 1633.7 m

Month	Mean sta. press. (mbar)	Mean daily temp. (°C)	Mean daily temp. range (°C)	Temp. extremes		Relative humidity (%)	Mean precip. (mm)	Max. precip. (24 h)
				highest (°C)	lowest (°C)			
Jan.		8.5	15.8	21.1	− 5.0	71	11.9	37.3
Feb.		9.7	14.4	23.3	− 6.7	71	39.5	33.3
Mar.		13.3	15.4	26.1	− 1.7	66	35.8	31.4
Apr.		15.9	13.3	29.0	2.8	68	63.1	45.0
May		18.3	10.7	29.5	6.5	78	129.2	88.6
June	no	19.8	6.8	30.5	11.7	87	235.6	71.9
July	data	21.2	6.6	30.0	12.2	87	313.2	88.1
Aug.		20.4	6.8	29.0	13.3	86	282.0	81.1
Sept.		16.8	7.7	29.0	9.5	84	161.4	58.0
Oct.		19.9	9.9	28.3	4.0	81	157.7	100.8
Nov.		12.8	13.2	24.5	− 2.0	78	41.3	71.4
Dec.		9.5	15.2	23.0	− 4.5	71	22.1	71.6
Annual		15.4	11.3	30.5	− 6.7	77	1497.8	100.8

Month	Mean evap. (mm)	Number of days			Number of cloudy days	Mean sunshine (h)	Most freq. wind direction	Mean wind speed (m/sec)
		precip.	thunder-storm	fog				
Jan.		2.5						
Feb.		7.1						
Mar.		6.8						
Apr.		11.6						
May		15.4						
June	no	23.0						
July	data	26.7			no data			
Aug.		24.9						
Sept.		18.1						
Oct.		13.9						
Nov.		5.6						
Dec.		2.9						
Annual		158.5						

TABLE LXI

CLIMATIC TABLE FOR KWEILIN
Latitude 25°15′N, longitude 110°10′E, elevation 166.7 m

Month	Mean sta. press. (mbar)	Mean daily temp.[1] (°C)	Mean daily temp. range (°C)	Temp. extremes		Relative humidity (%)	Mean precip.[1] (mm)	Max. precip. (24 h)	Max. snow cover (cm)
				highest (°C)	lowest (°C)				
Jan.		9.2	8.5	31.2	− 5.0	77	50.8	40.6	0.0
Feb.		9.7	7.6	31.3	− 3.3	79	79.1	53.3	0.0
Mar.		13.4	7.7	32.8	− 1.0	82	161.3	60.3	0.0
Apr.		19.0	9.2	34.9	2.9	80	222.6	104.1	0.0
May		23.7	9.4	38.0	10.2	84	358.8	162.3	0.0
June	no	23.7	9.4	39.4	15.0	82	370.4	159.5	0.0
July	data	28.4	10.1	39.7	17.5	82	235.7	190.3	0.0
Aug.		27.9	10.8	39.3	14.5	81	199.8	121.0	0.0
Sept.		26.5	12.2	39.4	13.5	70	100.6	101.2	0.0
Oct.		22.0	11.5	39.0	6.0	70	87.3	87.7	0.0
Nov.		15.6	9.9	32.9	2.0	74	53.2	54.4	0.0
Dec.		11.1	10.4	27.3	− 2.4	74	47.1	38.6	0.0
Annual		19.4	9.7	39.7	− 5.0	78	1966.1	190.3	0.0

Month	Mean evap. (mm)	Number of days			Number of cloudy days	Mean sunshine (h)	Most freq. wind direction	Mean wind speed (m/sec)	Number of days sand-storm
		precip.	thunder-storm	fog					
Jan.	67.0	11.6	0.5	8.8	22.1	79.4	N	2.6	0.0
Feb.	73.5	13.8	1.6	5.8	22.8	40.0	N	2.2	0.0
Mar.	87.0	19.1	5.1	7.0	24.8	57.3	N	2.5	0.0
Apr.	115.0	18.6	8.0	3.8	21.4	104.3	N	2.1	0.0
May	136.0	19.7	9.8	3.5	21.0	116.9	N	1.7	0.0
June	190.0	18.5	8.2	1.7	17.5	149.4	N	1.6	0.0
July	198.5	19.4	13.5	1.5	16.8	186.2	N, S	1.5	0.0
Aug.	182.5	15.5	13.9	3.3	14.1	217.6	N	1.3	0.0
Sept.	191.5	7.8	3.5	3.3	11.4	206.0	N	2.0	0.0
Oct.	176.0	7.9	1.1	4.2	13.5	149.5	N	2.7	0.0
Nov.	85.0	10.8	0.3	4.8	14.9	126.6	N	3.0	0.0
Dec.	75.0	11.1	0.3	8.0	16.1	135.1	N	3.0	0.0
Annual	1577.0	173.6	65.7	55.8	216.3	1568.3	N	2.2	0.0

[1] Mean daily temperature: 1935–1943, 1949–1950; mean precipitation: 1935–1950.

TABLE LXII

CLIMATIC TABLE FOR TAIPEI
Latitude 25°02′N, longitude 121°31′E, elevation 8.0 m

Month	M.S.L. press. (mbar)	Mean daily temp.[1] (°C)	Mean daily temp. range (°C)	Temp. extremes highest (°C)	Temp. extremes lowest (°C)	Relative humidity (%)	Mean precip.[1] (mm)	Max. precip. (24 h)
Jan.	1021.9	15.2	6.8	28.8	2.6	84	91	62.9
Feb.	1019.5	14.8	6.6	31.2	− 0.2	84	147	58.6
Mar.	1017.9	16.9	7.0	32.6	1.4	84	164	80.4
Apr.	1013.7	20.6	7.7	34.8	7.5	83	182	175.8
May	1009.7	24.1	7.9	36.5	10.0	82	205	168.6
June	1006.7	26.6	8.5	37.1	15.6	81	322	199.2
July	1006.1	28.2	8.6	38.6	19.5	78	269	358.9
Aug.	1005.1	27.9	8.7	37.7	18.9	78	266	325.8
Sept.	1009.3	26.2	8.4	36.4	13.5	80	189	293.1
Oct.	1015.0	23.0	7.4	36.1	10.8	81	117	198.7
Nov.	1018.6	19.8	6.9	33.6	1.1	81	71	57.2
Dec.	1020.9	16.8	6.8	31.5	1.8	83	77	79.3
Annual	1013.8	21.7	7.6	38.6	− 0.2	82	2100	358.9

Month	Mean evap. (mm)	Number of days precip.	Number of days thunder-storm	Number of days fog	Number of cloudy days	Mean sunshine (h)	Most freq. wind direction	Mean wind speed (m/sec)
Jan.		16.7	0.5	1.3	18.8	85.5	E	3.4
Feb.		16.6	1.0	1.9	19.8	74.1	E	3.3
Mar.		17.6	2.7	1.8	21.0	89.6	E	3.5
Apr.		15.1	3.5	1.1	17.5	112.1	E	3.1
May		16.3	4.3	1.0	17.1	138.0	E	2.9
June	no	15.6	8.6	1.3	15.6	168.1	E	2.2
July	data	14.0	10.3	0.6	9.3	225.2	E	2.5
Aug.		14.8	8.9	0.4	8.7	217.7	E	2.8
Sept.		14.0	4.5	0.4	9.7	191.2	E	3.1
Oct.		14.9	0.8	0.6	14.3	143.2	E	3.7
Nov.		14.9	0.2	1.1	15.9	107.3	E	3.9
Dec.		16.3	0.2	1.5	18.8	92.2	E	3.6
Annual		186.9	45.5	13.4	186.5	1644.2	E	3.2

[1] Mean daily temperature: 1897–1940; mean precipitation: 1889–1940.

TABLE LXIII

CLIMATIC TABLE FOR KUNMING
Latitude 25°02′N, longitude 102°43′E, elevation 1893.3 m

Month	Mean sta. press. (mbar)	Mean daily temp.[1] (°C)	Mean daily temp. range (°C)	Temp. extremes		Relative humidity (%)	Mean precip.[1] (mm)	Max. precip. (24 h)	Max. snow cover (cm)
				highest (°C)	lowest (°C)				
Jan.		9.6	14.0	26.5	− 5.4	64	3.4	12.8	5.0
Feb.		11.0	13.3	28.5	− 3.1	61	17.9	31.4	0.0
Mar.		14.5	14.1	29.0	− 1.4	56	21.2	39.5	0.0
Apr.		17.7	12.8	32.1	0.7	58	30.5	23.8	0.0
May		19.7	11.1	33.0	7.6	85	98.8	69.5	0.0
June	no	19.8	8.1	31.0	9.4	76	192.0	87.2	0.0
July	data	20.2	8.0	30.5	13.0	79	213.7	100.1	0.0
Aug.		19.9	7.9	30.5	12.0	79	220.2	105.0	0.0
Sept.		18.4	8.6	30.0	7.1	77	160.9	101.7	0.0
Oct.		15.6	9.0	28.5	1.5	77	94.6	77.4	0.0
Nov.		12.5	11.2	28.5	0.7	73	31.0	43.9	0.0
Dec.		9.8	12.9	26.0	− 2.9	70	10.5	15.8	0.0
Annual		15.7	10.8	33.0	− 5.4	70	1094.6	105.0	5.0

Month	Mean evap. (mm)	Number of days			Number of cloudy days	Mean sunshine (h)	Most freq. wind direction	Mean wind speed (m/sec)	Number of days sand-storm
		precip.	thunder-storm	fog					
Jan.	144.0	1.3	0.3	0.3	3.1	251.5	SW	2.8	0.0
Feb.	133.0	4.1	1.4	0.5	4.6	234.4	SW	3.1	0.0
Mar.	231.5	3.9	1.5	0.5	2.4	243.7	SW	3.2	0.0
Apr.	260.5	6.0	3.6	0.7	4.3	237.5	SW	3.3	0.0
May	241.5	12.2	4.6	0.0	8.7	193.7	SW	2.8	0.0
June	180.0	18.1	6.8	0.0	19.4	107.0	SSW	2.1	0.0
July	151.5	20.2	8.6	0.2	23.7	101.7	S	1.9	0.0
Aug.	153.5	13.9	7.7	0.0	19.5	130.7	NE	1.5	0.0
Sept.	141.5	15.1	2.4	0.0	16.1	118.1	E	1.9	0.0
Oct.	113.5	12.6	1.4	0.0	14.4	167.5	S	1.9	0.0
Nov.	134.0	6.4	0.5	0.0	8.1	180.7	SW	2.4	0.0
Dec.	106.5	2.6	0.1	0.6	4.9	202.8	SW	2.6	0.0
Annual	1991.0	121.4	39.6	2.8	129.2	2169.5	SW	2.4	0.0

[1] Mean daily temperature: 1928–1938, 1946–1952; mean precipitation: 1929–1938, 1946–1952.

TABLE LXIV

CLIMATIC TABLE FOR AMOY
Latitude 24°27′N, longitude 118°04′E, elevation 40.6 m

Month	M.S.L. press. (mbar)	Mean daily temp.[1] (°C)	Mean daily temp. range (°C)	Temp. extremes		Relative humidity (%)	Mean precip.[1] (mm)	Max. precip. (24 h)	Max. snow cover (cm)
				highest (°C)	lowest (°C)				
Jan.	1022.9	14.8	5.9	26.7	3.9	77	35.6	81.0	0.0
Feb.	1021.1	13.5	5.3	26.1	3.9	80	71.3	88.9	0.0
Mar.	1018.9	15.3	5.4	27.8	6.0	89	92.3	80.3	0.0
Apr.	1014.6	19.4	5.2	30.5	8.5	81	130.3	205.5	0.0
May	1010.2	23.4	5.0	32.5	13.9	81	170.7	183.6	0.0
June	1006.9	27.2	4.5	34.5	16.1	82	177.5	150.0	0.0
July	1005.4	29.0	4.9	36.1	16.7	79	132.2	127.0	0.0
Aug.	1004.7	29.0	4.9	37.9	23.0	79	164.5	128.0	0.0
Sept.	1009.9	28.1	5.3	37.5	18.3	74	107.7	159.4	0.0
Oct.	1015.9	25.0	6.4	35.0	13.3	70	37.5	190.7	0.0
Nov.	1019.7	20.8	6.3	32.5	6.7	72	32.3	75.5	0.0
Dec.	1021.7	16.6	6.0	27.8	5.0	75	33.5	36.9	0.0
Annual	1014.3	21.8	5.4	37.9	3.9	77	1185.6	205.5	0.0

Month	Mean evap. (mm)	Number of days			Number of cloudy days	Mean sunshine (h)	Most freq. wind direction	Mean wind speed (m/sec)	Number of days sand-storm
		precip.	thunder-storm	fog					
Jan.	55.7	7.2	0.0	0.3	17.6	125.3	ENE	3.2	0.0
Feb.	50.0	10.9	0.3	1.0	18.0	114.1	ENE	3.2	0.0
Mar.	77.7	12.0	1.7	2.0	22.7	84.5	ENE	3.1	0.0
Apr.	58.2	12.1	5.3	3.7	22.7	101.2	ENE	2.9	0.0
May	87.4	13.4	7.7	2.3	19.7	154.7	ESE	2.7	0.0
June	131.4	13.5	4.7	0.7	19.7	172.1	SW	2.8	0.0
July	189.8	9.6	8.3	0.0	14.7	253.6	SW	2.8	0.0
Aug.	205.6	10.7	8.7	0.0	10.0	243.8	SW	3.0	0.0
Sept.	186.8	7.5	5.7	0.0	16.0	189.1	SE	3.4	0.0
Oct.	157.5	3.6	1.0	0.0	9.0	208.4	ENE	3.8	0.0
Nov.	83.1	4.2	0.3	0.3	11.3	192.6	ENE	3.9	0.0
Dec.	56.8	5.8	0.0	1.7	13.3	149.1	ENE	3.4	0.0
Annual	1340.0	110.5	43.7	12.0	194.7	1988.6	ENE	3.2	0.0

[1] Mean daily temperature: 1891–1915; mean precipitation: 1910–1925.

TABLE LXV

CLIMATIC TABLE FOR TAINAN
Latitude 23°00′N, longitude 120°13′E, elevation 12.7 m

Month	M.S.L. press. (mbar)	Mean daily temp.[1] (°C)	Mean daily temp. range (°C)	Temp. extremes		Relative humidity (%)	Mean precip.[1] (mm)	Max. precip. (24 h)
				highest (°C)	lowest (°C)			
Jan.	1017.9	17.0	11.0	32.4	2.6	79	19.7	101.1
Feb.	1016.3	17.0	11.2	32.4	2.4	79	35.7	100.0
Mar.	1015.3	19.6	10.7	35.4	5.1	79	47.6	86.9
Apr.	1012.3	23.3	10.0	34.4	9.9	79	67.6	181.8
May	1009.0	26.2	9.0	36.4	14.7	81	170.8	252.3
June	1006.9	27.4	7.7	35.7	18.9	84	369.0	238.1
July	1005.7	27.8	7.9	36.9	21.1	83	431.3	397.8
Aug.	1004.9	27.3	7.5	36.6	21.2	85	448.2	384.9
Sept.	1008.2	27.1	8.6	36.6	15.4	82	156.0	382.1
Oct.	1012.1	24.8	10.2	34.7	12.6	76	33.4	79.4
Nov.	1015.0	21.6	10.8	35.2	2.9	78	15.9	43.2
Dec.	1017.0	18.5	10.8	32.0	4.3	79	15.9	54.0
Annual	1011.7	23.1	9.6	36.9	2.4	81	1811.1	397.8

Month	Mean evap. (mm)	Number of days			Number of cloudy days	Mean sunshine (h)	Most freq. wind direction	Mean wind speed (m/sec)
		precip.	thunder-storm	fog				
Jan.		4.9	0.3	2.4	7.3	195.9	N	3.9
Feb.		5.5	0.7	1.8	8.5	181.0	N	3.9
Mar.		7.0	1.5	1.6	9.3	198.0	N	3.6
Apr.		7.3	2.3	0.7	6.8	209.0	N	2.9
May		10.4	4.3	0.2	10.2	230.2	N	2.5
June	no	15.2	6.7	0.1	11.2	228.9	SE	2.6
July	data	16.9	11.5	0.1	10.7	236.1	SE	2.7
Aug.		19.0	10.9	0.1	11.8	214.0	SE	2.7
Sept.		10.8	7.7	0.3	6.0	241.6	N	2.5
Oct.		4.2	1.4	1.4	5.1	244.3	N	2.7
Nov.		3.3	0.1	1.5	5.2	212.2	N	3.2
Dec.		3.8	0.1	1.9	8.0	198.2	N	3.6
Annual		108.1	47.4	12.0	103.4	2588.1	N	3.1

[1] Mean daily temperature: 1892–1940; mean precipitation: 1892–1940.

TABLE LXVI

CLIMATIC TABLE FOR CANTON
Latitude 23°00′N, longitude 113°13′E, elevation 18.0 m

Month	M.S.L. press. (mbar)	Mean daily temp.[1] (°C)	Mean daily temp. range (°C)	Temp. extremes		Relative humidity (%)	Mean precip.[1] (mm)	Max. precip. (24 h)	Max. snow cover (cm)
				highest (°C)	lowest (°C)				
Jan.	1022.9	13.6	9.2	28.0	0.0	74	27.1	35.9	0.0
Feb.	1020.5	14.2	7.4	29.2	0.0	80	65.0	71.9	0.0
Mar.	1018.3	17.2	7.0	30.6	0.0	82	100.7	73.8	0.0
Apr.	1014.1	21.6	6.5	33.0	8.9	84	184.7	128.7	0.0
May	1010.1	25.6	7.1	35.7	10.6	83	256.1	143.5	0.0
June	1006.7	27.3	6.8	36.7	16.7	83	291.5	108.4	0.0
July	1005.7	28.8	7.8	37.2	21.0	80	264.4	182.5	0.0
Aug.	1005.3	28.2	7.8	37.7	20.4	81	248.5	125.7	0.0
Sept.	1010.2	27.2	8.3	37.6	13.7	77	149.1	131.4	0.0
Oct.	1016.3	24.0	9.8	36.0	10.0	71	48.5	76.6	0.0
Nov.	1020.1	19.7	10.1	32.0	1.1	76	50.5	132.5	0.0
Dec.	1022.2	15.7	9.6	29.0	− 0.3	71	34.1	56.3	0.0
Annual	1014.3	21.9	8.1	37.7	− 0.3	78	1720.1	182.5	0.0

Month	Mean evap. (mm)	Number of days			Number of cloudy days	Mean sunshine (h)	Most freq. wind direction	Mean wind speed (m/sec)	Number of days sand- storm
		precip.	thunder- storm	fog					
Jan.	127.5	7.3	0.2	6.3	14.0	125.0	N	2.1	0.0
Feb.	115.5	11.7	0.5	5.9	20.0	78.6	N	2.0	0.0
Mar.	97.0	14.2	2.0	9.5	29.0	82.9	N	1.8	0.0
Apr.	122.5	17.7	4.0	5.1	24.3	80.7	SE	1.8	0.0
May	138.5	17.3	5.6	3.5	21.0	154.1	SE	1.7	0.0
June	194.5	19.5	7.5	2.1	20.0	154.0	SE	1.6	0.0
July	186.5	16.7	8.5	1.3	16.7	214.9	SE	1.8	0.0
Aug.	162.5	16.9	8.3	2.3	14.3	214.4	E	1.7	0.0
Sept.	149.5	12.1	4.4	1.9	17.3	203.9	N	1.7	0.0
Oct.	197.0	6.4	0.7	4.5	11.7	209.0	N	1.9	0.0
Nov.	138.0	5.8	0.1	4.9	11.7	191.8	N	1.9	0.0
Dec.	123.5	7.1	0.0	6.9	15.0	158.0	N	1.9	0.0
Annual	1752.5	152.7	41.7	54.2	215.0	1867.3	N	1.8	0.0

[1] Mean daily temperature: 1912–1937, 1946–1952; mean precipitation: 1912–1952.

TABLE LXVII

CLIMATIC TABLE FOR NANNING
Latitude 22°48′N, longitude 108°18′E, elevation 74.9 m

Month	Mean sta. press. (mbar)	Mean daily temp.[1] (°C)	Mean daily temp. range (°C)	Temp. extremes highest (°C)	Temp. extremes lowest (°C)	Relative humidity (%)	Mean precip.[1] (mm)	Max. precip. (24 h)	Max. snow cover (cm)
Jan.		13.6	6.6	29.3	1.7	77	31.5	47.7	0.0
Feb.		14.4	6.1	32.7	2.3	77	55.2	43.1	0.0
Mar.		17.9	6.3	35.5	4.3	80	47.8	77.1	0.0
Apr.		22.5	6.6	36.0	8.0	78	79.2	52.0	0.0
May		26.7	6.6	36.7	14.8	79	167.0	102.3	0.0
June	no	28.0	6.0	37.4	16.0	81	214.2	118.6	0.0
July	data	28.5	5.9	38.3	19.5	81	216.9	114.2	0.0
Aug.		28.3	6.0	38.3	19.5	84	225.2	119.5	0.0
Sept.		27.6	6.7	38.8	13.5	76	109.0	128.0	0.0
Oct.		23.5	7.7	36.3	10.4	71	105.4	283.2	0.0
Nov.		19.7	7.9	32.2	4.4	72	36.8	36.3	0.0
Dec.		15.6	7.3	31.1	2.5	75	33.5	67.5	0.0
Annual		22.2	6.6	38.8	1.7	78	1321.8	283.2	0.0

Month	Mean evap. (mm)	Number of days precip.	Number of days thunder-storm	Number of days fog	Number of cloudy days	Mean sunshine (h)	Most freq. wind direction	Mean wind speed (m/sec)	Number of days sand-storm
Jan.	88.5	8.1	0.3	1.2	21.0	70.9	E	1.8	0.3
Feb.	82.0	10.7	0.3	1.6	21.7	67.3	E	2.0	0.0
Mar.	89.0	11.3	3.3	3.2	24.7	75.5	E	2.0	0.0
Apr.	135.5	11.4	5.6	1.2	17.3	109.1	E, SE	1.9	0.0
May	180.5	13.8	11.0	0.4	13.0	164.0	SE	1.9	0.0
June	222.0	15.0	12.3	0.4	18.3	178.4	SE	1.9	0.0
July	181.0	15.8	13.4	0.2	15.0	214.3	SE	2.0	0.0
Aug.	183.0	15.7	14.9	0.2	14.7	178.9	E	1.7	0.0
Sept.	185.5	8.6	6.3	1.0	11.3	181.1	E	1.4	0.0
Oct.	156.5	7.6	1.1	1.4	12.0	216.1	E	1.3	0.0
Nov.	107.0	5.7	0.0	0.6	12.0	153.0	E	1.4	0.0
Dec.	78.5	6.4	0.3	0.8	13.7	118.8	E	1.5	0.0
Annual	1689.0	130.3	67.7	12.2	194.7	1727.6	E	1.7	0.3

[1] Mean daily temperature: 1922–1939, 1941–1950; mean precipitation: 1907–1915, 1921–1950.

TABLE LXVIII

CLIMATIC TABLE FOR YULIN
Latitude 18°14′N, longitude 109°32′E, elevation 2.1 m

Month	Mean sta. press. (mbar)	Mean daily temp.[1] (°C)	Mean daily temp. range (°C)	Temp. extremes		Relative humidity (%)	Mean precip.[1] (mm)	Max. precip. (24 h)	Max. snow cover (cm)
				highest (°C)	lowest (°C)				
Jan.		21.4	7.9	30.0	9.0	75	10.8	3.3	0.0
Feb.		22.4	7.6	29.8	12.6	78	6.6	10.0	0.0
Mar.		24.4	7.1	31.6	16.1	79	20.6	2.6	0.0
Apr.		26.6	6.3	32.8	19.0	80	27.9	18.5	0.0
May		28.2	6.0	34.6	20.7	82	150.1	52.2	0.0
June	no	28.3	5.7	34.6	20.0	83	197.4	77.9	0.0
July	data	28.5	5.8	34.6	21.6	83	148.8	47.2	0.0
Aug.		28.0	6.1	34.6	22.7	84	188.8	43.5	0.0
Sept.		27.1	6.4	34.0	21.9	85	292.5	122.3	0.0
Oct.		25.9	7.5	34.0	15.5	80	190.0	119.9	0.0
Nov.		23.0	7.9	32.3	10.0	77	54.0	24.4	0.0
Dec.		21.4	8.0	30.2	6.6	73	42.5	37.3	0.0
Annual		25.5	6.9	34.6	6.6	80	1329.9	122.3	0.0

Month	Mean evap. (mm)	Number of days			Number of cloudy days	Mean sunshine (h)	Most freq. wind direction	Mean wind speed (m/sec)	Number of days sand-storm
		precip.	thunder-storm	fog					
Jan.		3.8	0.0	0.0	12.0	210.3	NW	2.5	0.0
Feb.		2.8	0.0	0.0	11.3	201.0	NW	2.3	0.0
Mar.		2.8	0.3	0.0	9.0	210.4	NW	2.3	0.0
Apr.		4.0	3.3	0.0	3.0	211.0	NW	2.1	0.0
May		10.2	10.7	0.0	17.7	230.0	NW	2.2	0.0
June	no	13.8	12.7	0.0	21.7	253.8	NW	2.2	0.0
July	data	14.2	11.3	0.0	16.7	286.2	SE	2.2	0.0
Aug.		16.4	14.7	0.0	20.3	234.8	SE	2.0	0.0
Sept.		18.6	13.7	0.0	20.7	203.2	NW	2.0	0.0
Oct.		12.4	9.0	0.0	13.7	242.6	SE	2.5	0.0
Nov.		6.8	1.0	0.0	12.3	210.4	NW	2.2	0.0
Dec.		6.8	0.0	0.0	14.0	196.3	NW	2.4	0.0
Annual		111.4	76.7	0.0	180.3	2752.0	NW	2.2	0.0

[1] Mean daily temperature: 1951–1955; mean precipitation: 1951–1953.

TABLE LXIX

CONVERSION TABLES

Temperatures		Rainfall and evap.		Velocity		
°C	°F	mm	inch	m/sec.	miles/h	knots
40	104.0	20	0.79	10	22.4	19.4
35	95.0	40	1.57	20	44.7	38.9
30	86.0	60	2.36	30	67.1	58.3
25	77.0	80	3.15	40	89.5	77.7
20	68.0	100	3.94	50	111.8	97.1
15	59.0	150	5.91	60	134.2	116.6
10	50.0	200	7.87	70	156.6	136.0
5	41.0	300	11.81	80	179.0	155.4
0	32.0	400	15.75	90	201.3	174.8
− 5	23.0	500	19.69	100	223.7	194.3
−10	14.0	600	23.62	1 m/sec = 2.237 miles/h = 1.943 knots		
−15	5.0	700	27.56	Distance		
−20	− 4.0	800	31.50	km		miles
−25	−13.0	900	35.43	1		0.6
−30	−22.0	1000	39.37	10		6
−35	−31.0	1500	59.06	100		60
−40	−40.0	2000	78.74	500		311
°F = (9/5)°C + 32		1 mm = 0.0394 inch		1 km = 0.621 statute mile		

Chapter 2

Climate of Japan

H. ARAKAWA and S. TAGA

Introduction

The climate of Japan is characterized mainly by the monsoon in winter, the Bai-u (rainy season) in early summer, and the typhoon in autumn. Almost every year tremendous losses of life and property are caused by meteorological disasters which are caused, more or less, by unusual climatic developments.

This paper will give an explication of these developments.

History of the development of climatology in Japan

Climatological studies in Japan started in the second half of the nineteenth century together with the development of the weather and climate observation net. In 1931, OKADA published his voluminous *Climate of Japan* which made a great contribution to the study of the climatic state of Japan.

Since that time, marked progress has been realized in Japan in the field of climatology. The main subjects studied in this period were micro- and local climatology, climatic division, air mass climatology and the physical explanation of seasonal phenomena.

The outbreak of World War II made it impossible to continue academic studies of climatology, but synoptic climatology was born as a by-product of war-time services.

After the War, in 1958, *Climate of Japan* was completed in cooperation with many climatologists and meteorologists under the supervision of WADACHI. This is the most up-to-date edition dealing with the climate of Japan. It contains the largest amount of collected data and illustrations on this topic now available.

The most active and important studies after the war were those dealing with dynamic climatology, local and meso-climatology (including topographic influence upon the temperature and rainfall distribution), change of climate and climatic classification.

Dynamic climatology

During the War, air mass analysis was commenced by ARAKAWA (1948), resulting in the development of dynamic climatology in Japan. TAKAHASHI (1955) explained the successive transition of seasons in terms of generation and movement of principal fronts, the passage of cyclones and anticyclones, etc. These papers are summarized in his eminent work *Dynamic Climatology*.

The completion of systematic upper air observations made it possible to study the synoptic situation over a broad area of the atmosphere. MURAKAMI (1951) clarified the double jet stream in the Bai-u season, one of which flows along the southern side of Japan, the other from Siberia or China to Kamchatka. The northern branch tends to form the blocking anticyclone over the Sea of Okhotsk and the cut-off low over Manchuria. The anticyclone over the Sea of Okhotsk was explained as one formed as a result of blocking action. The season ended simultaneously with the disappearance of the double jet stream system.

Local and meso-climatology

Meso-scale studies of climate have rapidly developed and many local discontinuity lines which are formed thermally or dynamically have been discovered in Japan. Those lines are named after the places where they are formed. Among these lines, the "Hokuriku discontinuity line", which causes heavy snowfall, has systematically been studied by the Japanese Meteorological Agency.

Climatic change

Of the studies of climate in Japan after the War, the study of climatic change was one of the most remarkable. Scientific symposiums on this subject were held several times and large numbers of studies covering a wide range of fields were made. Among these were small scale studies dealing with climatic fluctuations involving a few decades, as well as large scale studies involving hundreds of thousands of years. Concerning climatic changes in the past, the long rainfall records (1770–1907) discovered at Seoul, Korea, enabled ARAKAWA (1956) to demonstrate the remarkable dry period from the 1880's to the 1910's One recent trend of the studies is to investigate the climatic fluctuations of the past in relation to the climate of the Far East or of the Northern Hemisphere and to deduce a working theory from these fluctuations. The work by ARAKAWA (1957) and by YAMA-MOTO (1961) are worth noting in this respect. Yamamoto in particular has eagerly studied the climatic trends involving a few or more decades. He has made an example of the relationship between fluctuations of air temperature and sun-spots.

Climatic classification

For the division of climate, as is widely accepted, we have two opposing principles. One is empirical, environmental or distributional. The other is based on genetic or dynamic concepts. KIRA (1945) and KAWAKITA (1949) divided the climatic regions on the basis of the distribution of vegetation in eastern and southeastern Asia by using their thermal and humidity indices. Concerning classification of dynamical concepts, NAGAO (1961) classified the climate of Japan on a broad scale from the view-point of air-mass mixing and transformation. SUZUKI (1962) divided Japan using the polar front, regional differences of winter precipitation and concentration of heavy rainfall.

General character of the climate of Japan

Fig.1 shows the names and locations of the weather stations whose data were used in the tables.

Since Japan lies east of the Asian continent, the climate of Japan, excluding Hokkaido, is Cf and resembles that of the eastern coast of the North American continent. In order to compare the climatic data of five stations in Japan with those of five stations in the U.S.A., stations were chosen in such a way that their locations were evenly distributed from north to south and the latitude of each Japanese station was close to that of the corresponding American station (see Table I).

Fig.2 shows the average monthly mean temperatures of Japan in January and July, and Fig.3 shows the comparative seasonal changes of temperature from which the following climatic features can be derived: (*1*) in winter, especially midwinter, and spring, the air is much colder in Japan than along the eastern coast of the U.S.A. However, there is only a slight difference in temperature during midsummer and autumn.

The cold winter of Japan is due to the monsoonal system of wind which is highly developed in the Far East due to the vast continent of Asia.

(*2*) August is the hottest month in Japan, while in the U.S.A., July is the hottest month. The late arrival of the hottest season as mentioned above and the slow rise of temperature in spring as seen in Fig.3, demonstrate that the climate in Japan has more of a marine character than America. This, of course, is because Japan is an island.

The marine characteristics of Japan's climate appear in various aspects. The relative humidity (Fig.4) is one example, being remarkably high throughout the year.

Fig.5 shows the seasonal change of precipitation. The amount of precipitation is gener-

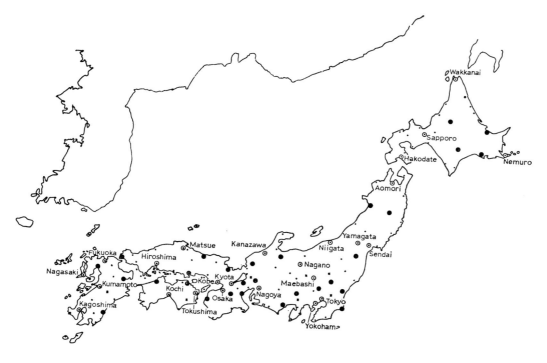

Fig.1. Map of stations.

Fig.2. Average monthly mean temperatures. A. January. B. July.

Fig.3. Average monthly temperatures for selected stations in Japan and the U.S.A.

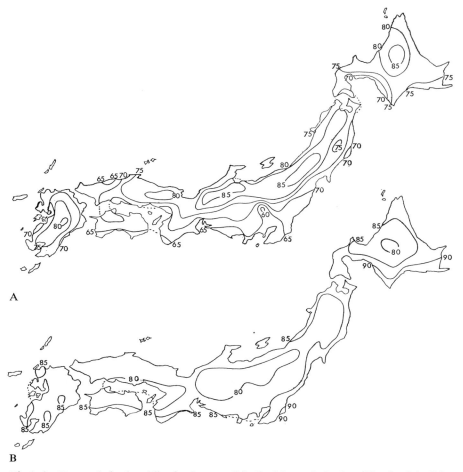

A

B

Fig.4. A. Mean relative humidity for January (%). B. Mean relative humidity for July (%).

ally higher in Japan than in the U.S.A. This figure, however, represents not only the marine climate of Japan, but differences of rain periods and their causes as well. As can be seen in Fig.5, two maxima of precipitation appear in the low latitudes of America. The first maximum in July is due to local thunderstorms and the second one, in September and October, is due to hurricanes. In the higher latitudes, as in the case of Wilmington, only one maximum, due to thundershowers, can be noticed. In the still higher latitudes, as in the case of Washington or Boston where precipitation is brought about mostly by

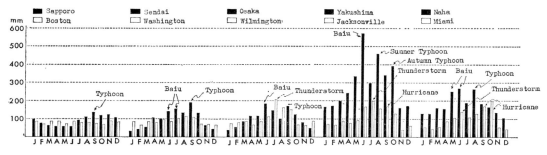

Fig.5. Average monthly precipitation for selected stations in Japan and the U.S.A.

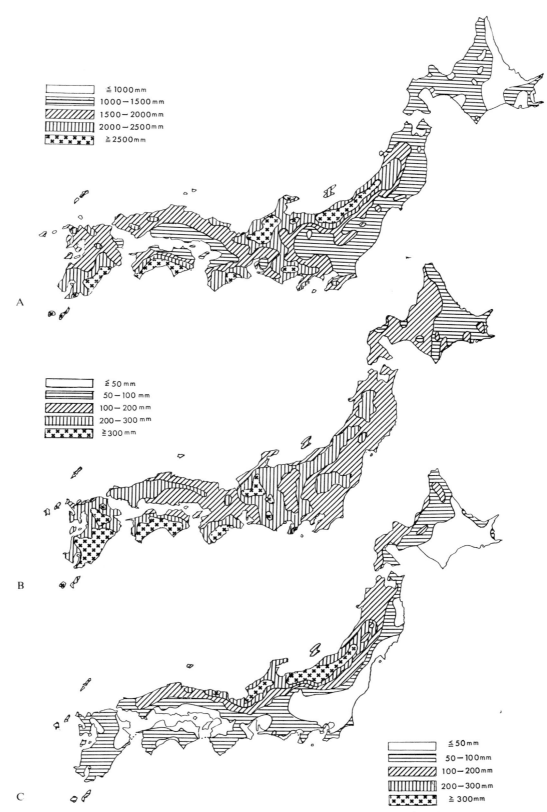

Fig.6. A. Precipitation in January. B. Precipitation in July. C. Annual precipitation.

TABLE I

STATIONS USED FOR COMPARISON OF CLIMATE

Station		Latitude	Longitude	Elevation (m)
Japan	Sapporo	43°03′N	141°20′E	18
U.S.A.	Boston, Mass.	42°22′N	71°01′W	5
Japan	Sendai	38°16′N	140°54′E	40
U.S.A.	Washington, D.C.	38°54′N	77°03′W	22
Japan	Osaka	34°39′N	135°32′E	8
U.S.A.	Wilmington, N.C.	34°16′N	77°55′W	9
Japan	Yakushima	30°27′N	130°30′E	15
U.S.A.	Jacksonville, Fla.	30°20′N	81°39′W	5
Japan	Naha, Ryukyu	26°12′N	127°39′E	30
U.S.A.	Miami, Fla.	25°47′N	80°11′W	2

frontal activity, the amount of precipitation is evenly distributed without conspicuous maxima.

In Japan, the seasonal change of precipitation is quite different from that of the U.S.A. In the case of Naha, two maxima are seen. The first maximum in June is due to the Bai-u. The second one in August is due to the typhoons. At Yakushima a similar condition exists, although a third maximum appears which is due to typhoons.

Two maxima, due to the Bai-u and typhoons, are also to be noticed at Osaka and Sendai where precipitation due to typhoons is not separated into two maxima. In Hokkaido there is practically no Bai-u season. Hence, at Sapporo, a Bai-u maximum does not appear. However, in November there is another maximum due to frontal disturbances. Fig.6 shows the average annual and monthly precipitation for the months of January and July.

To recapitulate, the climate of Japan is generally one of a marine character, but it is under the strong influence of the Asiatic winter monsoon. The large amount of precipitation is caused chiefly by Bai-u and typhoons.

Seasons in Japan

In a large part of Japan, six seasons can be recognized: winter, spring, Bai-u, summer, shurin (rainy season between summer and autumn) and autumn.

In winter a well-developed anticyclone persists on the Asian continent and the pattern of general circulation is settled. Summer is also a season of settled circulation as the anticyclone develops on the Pacific Ocean.

Between spring and summer there is a season of prolonged rainfall, i.e., Bai-u. Bai-u is caused by the stagnation of the polar front while it is moving from south to north.

Going from summer to winter the change is just the reverse. The rainy season between summer and autumn is called "Shurin" in Japanese. The stagnation of the front in this

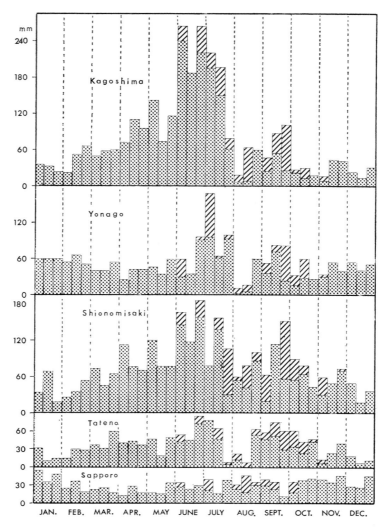

Fig.7. 10 day averages of precipitation (1951–1955). (After Saito et al., 1957.)

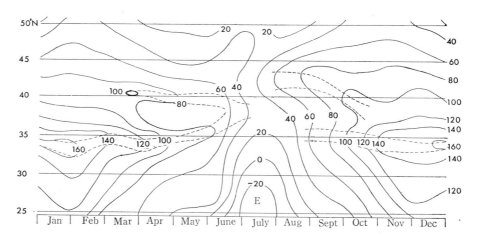

Fig.8. Monthly westerly wind speed (in knots) along the meridian 140° E at the level of highest speeds.

season is quite analogous to that of the Bai-u season. Shurin is a definite season, except at Kagoshima.

To make matters clear, the amount of precipitation due to typhoons is separated from the amount due to other causes. The result is shown in Fig.7. Both Bai-u and Shurin appear as peaks in this figure. Precipitation due to typhoons is shown by shaded areas and that due to other causes by scattered points. At Yonago the precipitation due to the winter monsoon is remarkable.

As seen in Fig.7, precipitation due to typhoons is distributed over the period June to October. The appearance of typhoons, however, is neither regular nor frequent. Therefore, it is not appropriate to regard this long period as one season.

The seasonal change on the ground corresponds to that of the general circulation in the upper atmosphere. Fig.8 shows the seasonal change of upper wind speeds (meridian 140° E) at the levels of highest speed. The jet stream is situated at its southernmost in winter and gradually moves northward in spring. In April the double jet streams appear. In summer the jet stream flows along its northernmost path at its slowest speed. From autumn to winter the path of the jet stream moves from north to south. The jet stream moves back and forth over the Japanese islands in a yearly cycle.

As is generally known, when the jet stream flows in the upper atmosphere, a frontal zone is liable to develop under it in the lower atmosphere. Accordingly, the annual march of the polar frontal zone over Japan causes the active interchange of air masses, thus characterizing the features of each season.

The climatic character of each season is represented most clearly in climatographs. Among the various types of climatographs, the type having the duration of sunshine as the ordinate and the daily maximum temperature as the abscissa has proven to be a powerful aid in discriminating seasons. The reason for this is that the march of seasons may be represented by the daily maximum temperature and the weather condition by the duration of sunshine.

Climatographs thus obtained can be classified into three types: the Pacific type, the Japan-Sea type and the Southeast-Coast-of-Hokkaido type. Fig.9 is typical of these climatographs.

(*a*) The Pacific type (Example: Murotomisaki). The change from winter to summer is similar to that from summer to winter. For instance, the Bai-u season corresponds to the Shurin season, both showing a diminution of sunshine.

(*b*) The Japan-Sea type (Example: Akita). The climatograph has a wide loop in the cold season which means that there is considerable reduction of sunshine in late autumn and early winter, possibly due to monsoon clouds.

(*c*) The Southeast-Coast-of-Hokkaido type (Example: Kushiro). The right part of the graph, that of the summer season, falls downward, which means that there is short duration of sunshine in this season owing to the dense sea fog.

According to SAITO et al. (1957), utilization of these climatographs at the beginning and end of each season can make it possible to fix the dates of the setting and passing of winter and summer in a normal year per day unit.

Table II shows the dates for winter and summer.

In spite of the difference of latitudes, boundaries between the adjoining four seasons are almost the same for several stations. The date of the beginning of Bai-u could not be determined definitely by use of a climatograph of any type, however, and it is even more

TABLE II

THE FIRST AND THE LAST DATES OF WINTER AND SUMMER AT SEVERAL STATIONS IN JAPAN
(After SAITO et al., 1957)

Station	Winter		Summer		Midsummer	
	the first date	the last date	the first date	the last date	the first date	the last date
Abashiri	Jan. 2	Feb. 19	July 14	Sept. 1	Aug. 3	Aug. 22
Sapporo	Dec. 30	Feb. 19	July 14	Aug. 31	Aug. 3	Aug. 22
Kanazawa	Jan. 2	Feb. 21	July 17	Aug. 31	Aug. 2	Aug. 23
Tokyo	Jan. 1	Feb. 21	July 14	Sept. 1	July 19	Aug. 24
Kagoshima	Dec. 29	Feb. 21	July 15	Sept. 2	July 21	Aug. 24

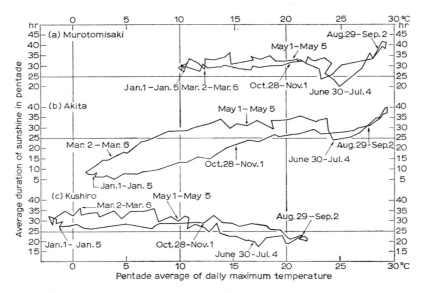

Fig.9. Climatographs showing three typical climatic types of Japan.

difficult to find the date of the passing of Shurin. Bai-u probably begins on or about June 15th in the southern part of Honshu and on June 25th, or thereabouts, in the northern part of the island. The date of the passing of Shurin is probably the first half of October or thereabouts.

Meteorological disasters and seasons

The number of damages each month from meteorological causes is shown in Table III. As seen in table, each of them has a tendency to arise in a certain peculiar season.

(*1*) Winter. Severe extratropical cyclones in this season cause ship disasters as well as damage along the coast due to wind waves. This indicates the violence of the winter northwesterly monsoon at the immediate rear of the cyclone.

(*2*) Spring. Since Japan is a country which rarely experiences deep snow, the rapid thawing of what does fall (owing to the rapid rise of temperature in spring) causes great flood damage in northern Japan.

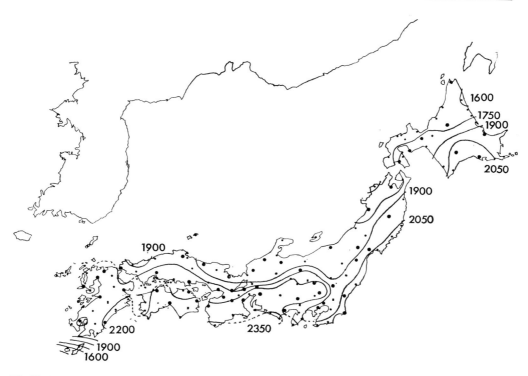

Fig.10. Mean annual duration of sunshine hours.

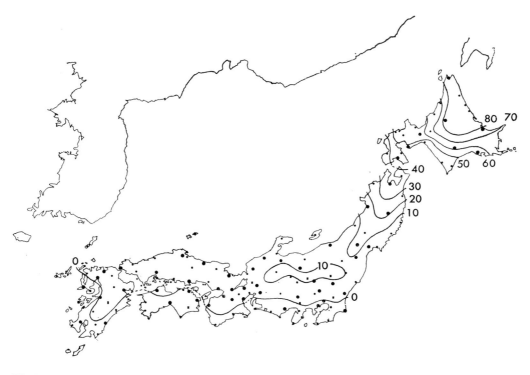

Fig.11. Number of days with the maximum temperature below 0° C.

TABLE III

SEASONAL DISTRIBUTION OF DAMAGE CAUSED BY METEOROLOGICAL PHENOMENA
(After SAITO et. al., 1957)

	J	F	M	A	M	J	J	A	S	O	N	D	Total	Notes
Typhoon						2	16	26	29	1	11		85	(1)
Heavy Bai-u				1	3	14	14	8	7	1	4		52	(1)
Cool summer						——5——							5	(2)
Flood due to thawing			2	8	4								14	(1)
Big fire	2	2	6	10	13	3			2	1	4	3	46	(3)
Extratropical cyclone	7	12	12	14					1	13	5	6	74	(1)
Drought						2	5	8	1		1		17	(4)
Frost damage				10	15	5							30	(4)

Notes:

(*1*) Frequency of disasters which caused losses of life and property in Japan during the period 1918–1947.

(*2*) Mean number of years in which the crops were exceptionally poor in the northern Japan during the period 1918–1947.

(*3*) Frequency of fires which burned more than 500 houses during the period 1923–1952.

(*4*) Frequency of serious damage during the period 1918–1947.

Large fires are also characteristic of the spring season. The rapid rise of temperature and the strong southwesterly winds in this season are two of the main causes of this phenomenon. Frost damage is peculiar to the inland area.

(*3*) Bai-u. Heavy rainfall is most serious in this season. In the late period of Bai-u, the southerly warm winds become strong along the southern side of the Bai-u front and bring abundant moisture from the Pacific Ocean. Under such circumstances, heavy showers which cause floods are often generated along the Bai-u front.

(*4*) Summer. The northern part of Japan has been subject to severe rice crop failures 10 times in the present century: 1902, 1905, 1913, 1931, 1934, 1935, 1941, 1945, 1953 and 1956. During these years some districts produced no rice crop at all. All of these bad crops resulted from unusual coolness and cloudiness in summer. These circumstances were caused by an abnormally prolonged stagnation of the Bai-u front, i.e., these cool summer resulted from an unusual delay in the passing of the Bai-u season. Another one of the frequent agricultural disasters in southern Japan is drought during the summer season.

(*5*) Shurin. In the period between June and November, some typhoons cause severe damage because of flood and storm surges. Typhoons reach their peak in September.

(*6*) Autumn. There are no remarkable meteorological damages peculiar to this season, except those due to occasional typhoons and cyclones.

Climatic tables of Japan

The data in the tables (see pp. 132–158) cover the new standard normal interval (1931–1960) chosen by the W.M.O.

Air pressure (at station level): Statistics of the air pressure at station level in Japan were begun in 1950 so that computations encompass data since 1951.

Wind: In 1949 it was decided to evaluate wind speed from the new empirical formula obtained from wind tunnel experiments. It was found that there was a gap in homogeneity between the wind data previous to 1949 and the data thereafter. Therefore, the wind norms used are from data since 1951.

References

ARAKAWA, H., 1948. *Climate of Japan.* Heibonsha, Tokyo, 157 pp. (Japanese).

ARAKAWA, H., 1956. On the secular variation of annual totals of rainfall at Seoul from 1770 to 1944. *Arch. Meteorol. Geophys. Bioklimatol. Ser B*, 7.: 205–211.

ARAKAWA, H., 1957. Climatic changes as revealed by the data from the Far East. *Weather*, 12(2): 46–51.

KAWAKITA, J., 1949. Numerical expression for the productivity of form-lands. *Social Geography*, 19: 6–10.

KIRA, T., 1945. A new classification of climate in southeastern Asia. *Rept. Agr. Lab., Kyoto Univ.*, pp. 1–24.

MURAKAMI, T., 1951. On the study of the change of the upper westerlies in the last stage of Bai-u season (rainy season in Japan). *J. Meteorol. Soc. Japan, Ser. II*, 29: 162–175.

MURAKAMI, T., 1959. The general circulation and water-vapour balance over the Far East during the rainy season. *Geophys. Mag.*, 29: 131–171.

NAGAO, T., 1961. Dynamical classification of climate based on the air-mass mixing and transformation. *Geograph. Rev. Japan*, 34: 307–320.

OKADA, T., 1931. *Climate of Japan.* Central Meteorological Observatory, Tokyo, 328 pp.

SAITO, R., ARAI, T., MIYAZAKI, M., KIKUCHIBARA, H. and KURIHARA, Y., 1957. The climate of Japan and her meteorological disasters. *Geophys. Mag.*, 28: 89–105.

SUZUKI, H., 1962. Classification of Japanese climates. *Geograph. Rev. Japan*, 35: 205–211.

TAKAHASHI, K., 1955. *Dynamic Climatology.* Iwanami Shoten, Tokyo, 316 pp. (Japanese).

WADACHI, K. (Editor), 1958. *Climate of Japan.* Tokyodo, Tokyo, 492 pp. (Japanese).

YAMAMOTO, T., 1961. Sunspot-climatic relationships in fluctuations of glaciers in the Alps and atmospheric precipitation in Korea. *J. Meteorol. Soc. Japan, Ser. II*, 39: 269–281.

YAMAMOTO, T., 1966. Synoptic aspects of climatic variation in the Far East with possible relation to sunspots. In: D. I. BLUMENSTOCK (Editor), *Pleistocene and Post-Pleistocene Climatic Variations in the Pacific Area.* Bishop Museum Press, Honolulu, pp.103–122.

TABLE IV

CLIMATIC TABLE FOR AOMORI
Latitude 40°49′N, longitude 140°47′E, elevation 3.6 m

Month	Mean stat. press. (mbar)	Mean daily temp. (°C)	Mean daily temp. range (°C)	Temp. extremes (°C) highest	Temp. extremes (°C) lowest	Mean vapor press. (mbar)	Mean precip. (mm)	Max. precip. (24 h)
Jan.	1013.2	− 2.7	7.2	13.5	−17.4	4.1	134.6	47.0
Feb.	1015.3	− 2.2	7.3	11.6	−18.7	4.3	107.5	81.8
Mar.	1014.4	0.4	7.8	20.0	−18.4	4.9	72.6	93.0
Apr.	1012.8	6.7	10.6	26.0	− 7.5	7.0	69.4	82.4
May	1010.6	12.3	11.2	29.2	− 1.4	10.6	66.8	86.5
June	1008.8	16.2	8.4	31.0	4.7	15.1	77.7	100.9
July	1007.4	20.4	7.3	34.2	7.2	20.4	111.5	112.1
Aug.	1008.8	22.3	8.4	36.0	9.5	22.7	139.4	187.9
Sept.	1012.1	18.0	10.0	35.9	3.0	17.1	148.2	106.7
Oct.	1017.3	11.9	10.8	30.5	− 2.2	11.1	110.3	74.9
Nov.	1018.9	6.0	8.7	23.7	− 8.8	7.4	120.9	107.9
Dec.	1015.2	0.2	6.2	21.1	−13.5	5.1	143.8	51.6
Annual	1012.9	9.1	25.0	36.0	−18.7	10.8	1299.8	187.9

Month	Mean evapor. (mm)	Number of days precip.	Number of days thunder- storms	Number of days gales	Number of days fog	Mean cloud- iness	Mean hours sun- shine	Wind most freq. direction	Wind mean speed (m/sec)
Jan.	37.8	25	0	15	1	8.8	56.7	SW	5.5
Feb.	40.8	20	0	14	1	8.8	75.6	SW	5.3
Mar.	65.9	16	–	13	1	7.8	141.7	SW	4.9
Apr.	95.2	11	0	15	1	6.6	200.8	SW	5.3
May	116.1	8	1	12	2	6.9	215.4	SW	4.5
June	113.9	9	2	8	3	7.4	196.8	SW	3.6
July	118.5	9	1	5	4	8.0	167.8	SW	3.3
Aug.	118.7	9	1	5	3	7.2	191.5	SW	3.2
Sept.	90.2	11	2	7	2	7.0	171.8	SW	3.5
Oct.	66.3	12	2	9	1	6.4	160.7	SW	3.6
Nov.	36.7	16	1	10	0	7.5	90.8	SW	4.2
Dec.	36.1	22	0	13	1	8.9	47.3	SW	5.1
Annual	936.1	166	9	127	19	7.6	1716.9	SW	4.3

TABLE V

CLIMATIC TABLE FOR FUKUOKA
Latitude 33°35′N, longitude 130°23′E, elevation 2.1 m

Month	Mean stat. press. (mbar)	Mean daily temp. (°C)	Mean daily temp. range (°C)	Temp. extremes (°C)		Mean vapor press. (mbar)	Mean precip. (mm)	Max. precip. (24 h)
				highest	lowest			
Jan.	1021.2	5.1	8.0	20.4	− 6.0	6.3	69.4	67.8
Feb.	1020.0	5.7	8.5	22.3	− 8.2	6.7	82.5	63.6
Mar.	1018.5	8.7	9.7	26.2	− 4.7	8.2	97.6	77.7
Apr.	1015.2	13.5	10.6	28.7	− 1.4	11.4	129.4	148.3
May	1011.4	17.8	10.3	30.2	1.4	15.3	127.4	93.8
June	1007.7	21.7	8.7	33.3	4.3	20.5	270.3	269.6
July	1007.4	26.3	7.6	36.3	13.8	27.1	252.9	182.9
Aug.	1008.1	26.8	8.3	36.7	15.4	27.7	170.8	153.4
Sept.	1011.2	22.8	8.6	35.0	7.9	22.4	244.0	189.0
Oct.	1016.9	16.9	10.7	32.4	0.4	14.9	101.7	171.9
Nov.	1020.4	12.2	10.6	27.7	− 2.1	10.7	79.5	40.8
Dec.	1022.1	7.6	8.7	22.1	− 5.4	7.7	77.8	36.0
Annual	1015.0	15.4	21.7	36.7	− 8.2	14.9	1703.4	269.6

Month	Mean evapor. (mm)	Number of days				Mean cloud-iness	Mean hours sun-shine	Wind	
		precip.	thunder-storms	gales	fog			most freq. direction	mean speed (m/sec)
Jan.	50.2	12	0	11	1	7.2	109.5	W	4.1
Feb.	54.6	10	0	8	1	7.3	124.4	N	3.8
Mar.	85.9	11	1	10	1	6.8	171.1	N	4.0
Apr.	109.2	10	1	9	1	6.4	188.2	N	3.6
May	134.5	11	1	6	1	6.9	205.4	N	3.3
June	134.1	12	1	5	1	7.7	176.7	N	3.1
July	165.3	11	4	3	0	7.2	204.7	N	3.1
Aug.	172.9	10	5	5	0	6.2	238.3	N	3.1
Sept.	117.3	12	2	7	0	6.8	168.7	N	3.1
Oct.	97.6	7	0	6	0	5.8	188.6	N	3.0
Nov.	65.1	8	0	7	1	5.9	159.8	N	2.9
Dec.	51.1	8	0	7	1	6.9	120.6	N	3.5
Annual	1237.8	122	16	83	7	6.8	2055.9	N	3.4

TABLE VI

CLIMATIC TABLE FOR HAKODATE
Latitude 41°49′N, longitude 140°45′E, elevation 33.3 m

Month	Mean stat. press. (mbar)	Mean daily temp. (°C)	Mean daily temp. range (°C)	Temp. extremes (°C) highest	Temp. extremes (°C) lowest	Mean vapor press. (mbar)	Mean precip. (mm)	Max. precip. (24 h)
Jan.	1009.2	− 4.1	7.1	8.8	−16.7	3.7	72.1	88.6
Feb.	1010.8	− 3.6	7.4	10.9	−17.9	3.8	56.6	49.9
Mar.	1010.6	0.0	7.3	14.0	−15.2	4.7	65.6	84.4
Apr.	1009.1	6.1	8.8	21.1	− 7.8	6.9	80.6	71.3
May	1007.0	11.0	9.2	24.7	− 0.9	10.0	85.8	57.0
June	1005.3	14.8	7.4	26.7	2.4	14.1	91.6	69.7
July	1004.8	19.4	6.6	31.1	8.5	19.5	133.5	114.1
Aug.	1005.8	21.6	7.1	31.4	11.3	22.0	128.7	176.0
Sept.	1008.4	17.2	8.4	28.7	4.5	16.4	179.7	122.1
Oct.	1012.9	11.3	9.3	26.6	− 2.2	10.6	116.0	98.9
Nov.	1013.8	4.6	8.1	20.1	− 9.7	6.7	90.3	146.8
Dec.	1011.3	− 1.3	6.6	14.8	−16.2	4.5	77.3	48.8
Annual	1009.1	8.1	25.7	31.4	−17.9	10.2	1177.9	176.0

Month	Mean evapor. (mm)	Numbers of days precip.	Numbers of days thunderstorms	Numbers of days gales	Numbers of days fog	Mean cloudiness	Mean hours sunshine	Wind most freq. direction	Wind mean speed (m/sec)
Jan.	x	13	0	15	0	7.2	108.9	WNW	4.6
Feb.	x	12	0	14	1	7.1	124.3	WNW	4.9
Mar.	x	12	–	16	1	6.9	164.7	WNW	4.8
Apr.	x	11	1	16	1	6.3	205.1	W	4.9
May	115.3	9	1	14	2	6.6	214.7	E	4.6
June	109.9	9	1	7	4	7.3	180.9	E	4.0
July	103.0	9	1	4	5	7.8	160.7	E	3.4
Aug.	119.7	10	2	3	1	7.1	183.9	E	3.3
Sept.	100.2	11	2	7	0	6.6	173.7	E	3.7
Oct.	81.4	10	2	10	0	5.7	179.6	WNW	4.0
Nov.	x	11	1	11	0	6.3	131.4	WNW	4.1
Dec.	x	12	0	13	1	7.4	93.9	WNW	4.5
Annual	x	128	8	129	16	6.9	1921.7	WNW	4.2

TABLE VII

CLIMATIC TABLE FOR HIROSHIMA
Latitude 34°22′N, longitude 132°26′E, elevation 29.1 m

Month	Mean stat. press. (mbar)	Mean daily temp. (°C)	Mean daily temp. range (°C)	Temp. extremes (°C)		Mean vapor press. (mbar)	Mean precip. (mm)	Max. precip. (24 h)
				highest	lowest			
Jan.	1017.0	4.2	9.0	18.8	− 8.3	5.8	45.1	58.1
Feb.	1016.2	4.7	9.4	19.2	− 8.3	6.0	69.8	66.2
Mar.	1014.9	7.6	9.7	22.4	− 7.2	7.4	106.4	75.2
Apr.	1012.2	12.7	11.2	25.0	− 1.4	10.5	157.7	122.4
May	1008.6	17.1	9.3	29.6	1.8	14.5	153.8	109.3
June	1004.9	21.0	8.4	30.9	6.6	19.7	249.4	159.9
July	1004.8	25.4	6.6	35.0	14.2	26.5	249.5	152.1
Aug.	1005.5	26.6	8.6	36.7	13.7	26.9	115.5	183.5
Sept.	1008.5	22.7	8.2	33.2	8.8	21.4	215.5	339.6
Oct.	1013.8	16.7	10.3	30.0	1.5	14.3	114.6	159.1
Nov.	1017.1	11.5	10.7	25.4	− 2.6	10.1	67.0	76.3
Dec.	1018.1	6.6	9.4	20.5	− 8.6	7.1	51.2	47.1
Annual	1011.8	14.7	22.4	36.7	− 8.6	14.2	1595.5	339.6

Month	Mean evapor. (mm)	Number of days				Mean cloud-iness	Mean hours sun-shine	Wind	
		precip.	thun-der-storms	gales	fog			most freq. direction	mean speed (m/sec)
Jan.	49.7	8	–	8	1	6.4	145.8	NNE	3.7
Feb.	54.1	8	0	8	2	6.4	153.0	NNE	3.8
Mar.	82.0	10	0	8	2	6.1	187.5	NNE	3.9
Apr.	108.3	10	1	6	3	6.1	204.6	NNE	3.6
May	132.4	11	1	4	2	6.4	222.1	NNE	3.1
June	128.7	12	1	2	2	7.4	191.1	NNE	2.9
July	158.0	13	3	2	3	6.9	211.0	NNE	2.7
Aug.	181.3	8	3	2	0	5.9	247.6	NNE	3.2
Sept.	119.6	12	2	4	1	6.7	177.5	NNE	3.6
Oct.	94.1	6	0	4	1	5.5	190.0	NNE	4.2
Nov.	63.2	5	0	5	1	5.2	176.7	NNE	4.1
Dec.	50.6	5	–	7	1	5.9	154.9	NNE	3.8
Annual	1222.1	107	10	58	19	6.2	2261.8	NNE	3.5

TABLE VIII

CLIMATIC TABLE FOR KAGOSHIMA
Latitude 31°34′N, longitude 130°33′E, elevation 4.8 m

Month	Mean stat. press. (mbar)	Mean daily temp. (°C)	Mean daily temp. range (°C)	Temp. extremes (°C) highest	lowest	Mean vapor press. (mbar)	Mean precip. (mm)	Max. precip. (24 h)
Jan.	1020.3	6.6	10.2	23.9	− 5.7	7.6	75.3	113.5
Feb.	1018.8	7.7	10.4	24.1	− 6.7	8.0	116.0	90.2
Mar.	1017.3	10.8	10.7	25.8	− 3.9	9.8	149.0	94.4
Apr.	1014.7	15.1	10.7	27.7	− 1.0	13.2	227.7	145.1
May	1011.0	19.0	9.8	31.1	3.9	17.2	249.3	155.9
June	1007.8	22.6	8.0	34.1	9.0	22.4	454.3	305.7
July	1008.0	26.8	7.7	36.6	15.9	28.3	343.1	233.8
Aug.	1007.7	27.1	8.3	37.0	16.8	28.2	220.2	216.8
Sept.	1010.2	24.4	9.1	34.2	9.8	24.1	213.3	174.0
Oct.	1015.3	18.9	10.7	32.2	2.6	16.6	120.3	166.9
Nov.	1019.0	14.0	11.3	28.6	− 1.5	12.5	89.5	98.1
Dec.	1021.1	9.0	11.0	24.4	− 5.5	9.0	79.1	169.2
Annual	1014.3	16.8	20.5	37.0	− 6.7	16.4	2337.1	305.7

Month	Mean evapor. (mm)	Number of days precip.	thunder-storms	gales	fog	Mean cloud-iness	Mean hours sun-shine	Wind most freq. direction	mean speed (m/sec)
Jan.	67.8	10	0	7	0	6.0	153.3	NW	4.1
Feb.	73.7	10	0	7	0	6.3	147.4	NW	4.1
Mar.	102.5	13	1	9	0	6.3	178.3	NW	4.2
Apr.	119.4	13	1	6	1	6.6	179.0	NW	4.1
May	139.6	14	1	6	0	7.3	189.6	NW	3.8
June	131.1	15	2	5	0	8.1	158.6	NW	3.6
July	179.2	13	3	5	0	7.0	221.7	NW	3.6
Aug.	195.4	11	4	5	0	6.0	243.6	NW	4.0
Sept.	148.0	11	3	5	0	6.2	200.9	NW	3.9
Oct.	124.7	8	1	5	0	5.3	197.7	NW	4.0
Nov.	85.9	7	0	5	0	5.1	179.6	NW	3.9
Dec.	69.3	8	1	6	0	5.3	170.6	NW	3.9
Annual	1436.6	133	17	69	4	6.3	2220.2	NW	3.9

TABLE IX

CLIMATIC TABLE FOR KANAZAWA
Latitude 36°33′N, longitude 136°39′E, elevation 27.0 m

Month	Mean stat. press. (mbar)	Mean daily temp. (°C)	Mean daily temp. range (°C)	Temp. extremes (°C)		Mean vapor press. (mbar)	Mean precip. (mm)	Max. precip. (24 h)
				highest	lowest			
Jan.	1014.8	2.5	6.1	21.2	− 9.7	5.8	309.2	71.6
Feb.	1015.2	2.5	6.7	23.6	− 9.4	5.6	190.7	61.4
Mar.	1014.3	5.5	8.5	25.0	− 8.3	6.6	172.9	69.4
Apr.	1012.3	11.0	10.4	29.9	− 1.6	9.2	163.8	71.8
May	1008.8	16.1	10.0	32.4	1.5	13.2	135.3	83.2
June	1005.2	20.2	8.4	34.5	6.8	18.7	166.9	146.8
July	1005.0	24.5	7.8	36.9	11.0	24.8	223.4	178.5
Aug.	1005.2	25.9	8.9	37.1	14.5	25.9	154.3	115.1
Sept.	1009.0	21.7	8.5	38.5	7.6	20.5	248.0	156.3
Oct.	1013.9	15.6	8.6	33.0	2.2	13.8	217.1	130.8
Nov.	1016.6	10.5	8.6	27.6	− 0.7	9.7	225.0	89.3
Dec.	1016.1	5.6	6.9	23.6	− 6.2	7.1	352.7	85.1
Annual	1011.4	13.5	23.4	38.5	− 9.7	13.4	2559.3	178.5

Month	Mean evapor. (mm)	Number of days				Mean cloud-iness	Mean hours sun-shine	Wind	
		precip.	thunder-storms	gales	fog			most freq. direction	mean speed (m/sec)
Jan.	34.8	24	2	14	0	8.7	62.3	ESE	4.7
Feb.	39.5	19	1	10	0	8.5	84.0	ESE	4.2
Mar.	65.2	18	1	10	0	7.5	142.4	ESE	4.1
Apr.	99.3	13	1	9	0	6.5	194.8	ESE	3.9
May	124.4	12	1	6	0	6.7	213.0	E	3.6
June	120.4	13	2	3	0	7.5	185.9	E	3.3
July	134.3	13	3	2	0	7.3	190.1	E	2.9
Aug.	154.9	10	3	2	0	6.0	240.8	E	3.0
Sept.	104.7	13	2	3	0	7.3	158.9	E	3.1
Oct.	72.5	13	1	4	0	6.8	150.9	ESE	3.3
Nov.	51.4	16	2	7	0	6.9	122.8	ESE	3.7
Dec.	39.3	23	2	13	0	8.3	70.4	ESE	4.3
Annual	1040.6	188	21	84	2	7.3	1816.4	ESE	3.7

TABLE X

CLIMATIC TABLE FOR KOBE
Latitude 34°41′N, longitude 135°11′E, elevation 58.1 m

Month	Mean stat. press. (mbar)	Mean daily temp. (°C)	Mean daily temp. range (°C)	Temp. extremes (°C) highest	lowest	Mean vapor press. (mbar)	Mean precip. (mm)	Max. precip. (24 h)
Jan.	1011.8	4.5	7.1	19.2	− 6.4	5.5	40.9	44.2
Feb.	1011.6	4.8	7 6	20 8	− 5.8	5.7	56.2	57.9
Mar.	1010.5	7.7	8.5	23.7	− 4.2	7.0	93.7	117.1
Apr.	1008.6	13.2	9.0	28.4	− 0.6	10.1	122.9	81.4
May	1005.1	18.0	8.6	31.9	3.9	13.9	122.7	112.3
June	1001.6	21.7	7.4	33.9	10.0	19.1	181.5	181.8
July	1001.7	26.0	6.9	36.8	15.0	25.4	184.0	270.4
Aug.	1002.5	27.1	7.6	37.6	16.1	25.9	135.9	143.8
Sept.	1005.1	23.5	7.4	34.3	10.5	21.0	175.3	199.4
Oct.	1009.7	17.6	7.8	29.6	5.3	14.0	104.5	262.8
Nov.	1012.8	12.5	7.8	26.0	0.5	9.9	75.9	76.3
Dec.	1013.2	7.4	7.3	22.1	− 4.3	6.9	43.6	44.5
Annual	1007.9	15.3	22.6	37.6	− 6.4	13.7	1336.9	270.4

Month	Mean evapor. (mm)	Number of days precip.	thunder- storms	gales	fog	Mean cloud- iness	Mean hours sun- shine	Wind most freq. direction	mean speed (m/sec)
Jan.	67.3	6	−	14	2	5.3	158.2	W	4.6
Feb.	64.7	7	0	12	1	5.9	148.8	WNW	4.3
Mar.	89.0	10	0	12	2	6.0	181.2	N	3.9
Apr.	120.2	10	1	11	1	6.1	197.8	ENE	4.0
May	147.7	11	1	11	1	6.7	218.5	ENE	4.1
June	138.6	12	1	7	1	7.6	180.7	ENE	3.9
July	174.3	11	3	7	1	7.0	209.0	WSW	3.8
Aug.	195.5	8	3	8	0	6.0	238.8	N	4.1
Sept.	138.3	11	2	8	0	6.8	174.5	N	3.9
Oct.	103.3	8	0	9	0	6.0	171.6	N	4.0
Nov.	75.5	6	0	8	1	5.2	160.9	N	3.7
Dec.	69.1	5	0	11	2	5.0	156.5	W	4.3
Annual	1383.4	105	10	115	12	6.1	2196.3	N	4.0

TABLE XI

CLIMATIC TABLE FOR KOCHI
Latitude 33°34′N, longitude 133°33′E, elevation 0.5 m

Month	Mean stat. press. (mbar)	Mean daily temp. (°C)	Mean daily temp. range (°C)	Temp. extremes (°C)		Mean vapor press. (mbar)	Mean precip. (mm)	Max. precip. (24 h)
				highest	lowest			
Jan.	1019.0	5.2	11.2	23.4	− 7.6	6.0	54.9	130.2
Feb.	1018.3	6.3	11.0	23.7	− 7.0	6.5	96.6	127.7
Mar.	1017.0	9.6	10.9	26.3	− 6.5	8.4	177.2	195.4
Apr.	1015.1	14.4	10.5	30.0	− 0.9	12.0	260.9	194.6
May	1011.8	18.5	9.6	31.2	3.8	16.2	278.9	239.1
June	1008.4	21.8	7.9	33.9	9.1	21.6	344.1	263.2
July	1008.5	25.7	7.2	37.1	14.6	28.0	368.6	291.3
Aug.	1009.0	26.3	8.3	37.3	15.9	28.1	343.6	364.3
Sept.	1011.4	23.5	8.9	35.9	11.3	23.2	350.4	370.8
Oct.	1015.9	18.0	10.4	32.2	2.5	15.5	183.9	211.8
Nov.	1019.2	12.9	11.2	27.8	− 1.9	11.0	107.5	116.2
Dec.	1020.3	7.8	11.4	23.5	− 6.6	7.5	79.7	165.8
Annual	1014.5	15.8	21.1	37.3	− 7.6	15.3	2646.3	370.8

Month	Mean evapor. (mm)	Number of days				Mean cloud-iness	Mean hours sun-shine	Wind	
		precip.	thun-der-storms	gales	fog			most freq. direction	mean speed (m/sec)
Jan.	58.4	6	0	5	0	4.5	193.9	W	2.5
Feb.	64.2	7	0	5	1	5.2	178.7	W	2.5
Mar.	89.6	10	1	4	1	5.8	200.1	W	2.6
Apr.	107.4	12	1	3	1	6.3	196.0	W	2.7
May	120.0	13	1	2	1	7.0	200.5	W	2.4
June	108.5	14	1	1	1	8.0	162.2	W	2.1
July	129.7	14	4	2	1	7.3	194.2	SSE	2.0
Aug.	146.9	12	5	2	1	6.4	222.1	WNW	2.3
Sept.	110.2	13	3	2	0	6.8	178.4	W	2.3
Oct.	91.5	8	1	1	0	5.7	183.6	W	2.2
Nov.	66.9	6	0	2	−	4.8	182.3	WNW	2.2
Dec.	57.0	4	0	2	0	4.3	190.5	W	2.5
Annual	1150.3	118	16	30	6	6.0	2282.5	W	2.3

TABLE XII

CLIMATIC TABLE FOR KUMAMOTO
Latitude 32°49′N, longitude 130°43′E, elevation 37.9 m

Month	Mean stat. press. (mbar)	Mean daily temp. (°C)	Mean daily temp. range (°C)	Temp. extremes (°C)		Mean vapor press. (mbar)	Mean precip. (mm)	Max. precip. (24 h)
				highest	lowest			
Jan.	1016.9	4.6	10.4	22.5	− 9.2	6.4	51.7	64.1
Feb.	1015.3	5.7	10.8	26.4	− 9.2	6.7	79.6	116.5
Mar.	1013.7	9.2	11.7	27.4	− 6.9	8.4	108.6	74.6
Apr.	1010.8	14.1	12.3	29.8	− 2.5	11.7	169.8	143.1
May	1007.0	18.6	11.7	32.8	1.3	15.6	186.2	146.7
June	1003.6	22.4	9.7	34.8	7.1	20.9	345.3	411.9
July	1003.7	26.5	8.3	38.8	14.3	27.5	330.8	480.5
Aug.	1003.9	27.0	9.6	37.1	15.3	27.2	167.2	220.0
Sept.	1006.6	23.4	10.0	35.8	6.7	22.5	198.7	136.9
Oct.	1012.1	17.4	12.0	32.8	0.5	15.0	95.4	125.3
Nov.	1015.7	12.3	12.3	28.9	− 3.8	10.9	64.3	93.9
Dec.	1017.8	7.1	11.2	24.6	− 7.9	7.7	71.7	67.8
Annual	1010.6	15.7	22.4	38.8	− 9.2	15.0	1869.2	480.5

Month	Mean evapor. (mm)	Number of days				Mean cloud-iness	Mean hours sun-shine	Wind	
		precip.	thunder. storms	gales	fog			most freq. direction	mean speed (m/sec)
Jan.	48.4	8	0	3	3	6.3	134.4	N	2.2
Feb.	59.0	8	0	2	3	6.3	142.3	N	2.3
Mar.	88.4	10	1	3	2	6.3	180.3	N	2.4
Apr.	109.2	12	1	2	3	6.4	188.5	N	2.3
May	127.4	13	1	1	2	7.0	200.0	SW	2.1
June	126.8	14	1	2	2	7.9	172.6	SW	2.2
July	155.2	14	5	1	2	7.3	201.5	SW	2.3
Aug.	171.6	9	5	3	2	6.2	236.8	SW	2.2
Sept.	123.4	11	2	2	2	6.6	183.5	N	2.1
Oct.	100.1	7	1	1	2	5.3	193.3	N	2.0
Nov.	67.3	6	0	0	3	5.2	172.5	N	1.9
Dec.	49.1	6	0	2	3	5.7	142.5	N	2 0
Annual	1225.7	116	17	22	28	6.4	2148.1	N	2.2

TABLE XIII

CLIMATIC TABLE FOR KYOTO
Latitude 35°01′N, longitude 135°44′E, elevation 40.9 m

Month	Mean stat. press. (mbar)	Mean daily temp. (°C)	Mean daily temp. range (°C)	Temp. extremes (°C)		Mean vapor press. (mbar)	Mean precip. (mm)	Max. precip. (24 h)
				highest	lowest			
Jan.	1014.1	3.3	9.5	19.9	− 8.4	5.8	51.2	59.2
Feb.	1013.9	3.8	9.7	21.2	− 8.8	5.9	65.4	62.4
Mar.	1012.9	6.9	11.0	25.7	− 6.5	7.1	107.8	101.0
Apr.	1011.0	12.5	12.4	29.3	− 2.8	10.0	140.7	104.9
May	1007.5	17.6	12.0	33.0	1.6	13.7	142.7	98.4
June	1003.9	21.7	10.3	34.6	6.0	19.0	232.8	281.6
July	1003.9	26.0	9.6	38.2	11.5	25.3	216.2	160.5
Aug.	1004.7	26.9	10.3	38.2	12.8	25.8	160.0	288.6
Sept.	1007.4	22.9	10.0	36.0	8.6	21.2	201.5	122.1
Oct.	1012.2	16.5	10.8	30.9	1.4	14.3	124.3	121.5
Nov.	1015.2	10.9	11.3	26.9	− 2.1	10.1	83.2	72.4
Dec.	1015.5	5.8	10.1	21.3	− 9.4	7.1	53.1	53.9
Annual	1010.2	14.6	23.6	38.2	− 9.4	13.8	1578.8	288.6

Month	Mean evapor. (mm)	Number of days				Mean cloud-iness	Mean hours sun-shine	Wind	
		precip.	thun-der-storms	gales	fog			most freq. direction	mean speed (m/sec)
Jan.	39.2	8	0	2	10	6.1	135.4	W	1.9
Feb.	42.6	7	0	2	7	6.5	131.4	N	2.2
Mar.	65.5	11	0	2	5	6.4	167.7	NW	2.3
Apr.	93.1	10	1	2	2	6.4	189.2	NNW	2.3
May	121.3	12	1	1	2	6.7	206.4	NE	2.2
June	118.0	13	2	1	1	7.8	165.4	NE	2.1
July	142.8	12	5	1	1	7.2	191.0	SSW	2.1
Aug.	154.1	8	6	1	0	6.2	220.6	ENE	2.2
Sept.	104.2	13	3	1	2	7.0	161.1	N	1.9
Oct.	72.7	8	0	1	5	6.2	161.3	NNW	1.7
Nov.	47.0	7	0	1	11	5.5	150.3	N	1.5
Dec.	37.1	6	0	1	15	5.7	133.0	W	1.7
Annual	1037.5	116	19	15	60	6.5	2012.5	NNW	2.0

TABLE XIV

CLIMATIC TABLE FOR MAEBASHI
Latitude 36°24′N, longitude 139°04′E, elevation 111.7 m

Month	Mean stat. press. (mbar)	Mean daily temp. (°C)	Mean daily temp. range (°C)	Temp. extremes (°C) highest	Temp. extremes (°C) lowest	Mean vapor press. (mbar)	Mean precip. (mm)	Max. precip. (24 h)
Jan.	1002.0	2.4	10.3	22.0	−11.8	4.2	20.9	27.9
Feb.	1002.7	2.9	10.2	24.6	− 8.9	4.3	33.2	55.3
Mar.	1002.1	6.1	10.5	26.3	− 7.8	5.5	48.5	59.9
Apr.	1001.6	11.6	11.0	32.4	− 3.1	8.6	77.0	99.7
May	998.9	16.4	10.8	36.1	0.3	12.9	98.7	119.8
June	995.9	20.4	8.8	35.4	6.0	18.1	165.8	111.5
July	995.8	24.4	7.8	37.4	13.9	24.4	198.3	164.7
Aug.	997.0	25.3	8.0	37.3	13.6	25.9	198.5	262.4
Sept.	999.7	21.2	7.9	36.0	8.4	20.4	195.6	357.4
Oct.	1003.9	15.3	9.1	33.0	0.6	13.2	138.1	148.0
Nov.	1005.5	10.1	10.1	26.6	− 3.5	8.3	47.5	71.5
Dec.	1003.7	5.1	9.3	21.9	− 7.4	5.3	24.3	64.5
Annual	1000.7	13.4	22.9	37.4	−11.8	12.6	1246.4	357.4

Month	Mean evapor. (mm)	Number of days precip.	Number of days thunder-storms	Number of days gales	Number of days fog	Mean cloud-iness	Mean hours sun-shine	Wind most freq. direction	Wind mean speed (m/sec)
Jan.	89.6	3	−	15	0	3.5	212.5	NNW	5.1
Feb.	93.6	5	0	13	0	4.4	195.0	NNW	4.9
Mar.	130.0	8	0	15	0	5.4	214.9	NNW	5.1
Apr.	150.5	9	1	11	0	6.2	206.6	NNW	4.6
May	155.7	12	2	6	1	6.9	209.4	NNW	3.7
June	133.1	14	3	3	1	8.3	150.0	ESE	3.1
July	146.9	14	7	2	1	8.2	163.9	ESE	2.5
Aug.	148.8	13	9	2	1	7.5	189.7	ESE	2.9
Sept.	104.0	14	2	4	1	7.8	135.8	NNW	2.9
Oct.	94.0	10	1	4	1	6.6	157.2	NNW	3.2
Nov.	89.4	6	−	8	1	4.9	176.1	NNW	4.1
Dec.	86.0	4	−	12	0	3.6	198.7	NNW	4.7
Annual	1421.8	110	24	96	9	6.1	2209.8	NNW	3.9

TABLE XV

CLIMATIC TABLE FOR MATSUE
Latitude 35°27′N, longitude 133°04′E, elevation 17.1 m

Month	Mean stat. press. (mbar)	Mean daily temp. (°C)	Mean daily temp. range (°C)	Temp. extremes (°C)		Mean vapor press. (mbar)	Mean precip. (mm)	Max. precip. (24 h)
				highest	lowest			
Jan.	1017.6	3.8	6.7	18.5	− 6.2	/	170.5	44.1
Feb.	1017.4	3.9	7.3	24.7	− 8.5	/	161.6	50.4
Mar.	1016.2	6.9	9.0	24.8	− 4.4	/	129.5	47.9
Apr.	1013.4	12.0	10.5	29.5	− 2.1	/	115.9	94.8
May	1009.8	16.6	9.7	31.5	3.4	/	132.1	76.4
June	1006.1	20.8	8.1	34.3	7.8	/	189.2	128.9
July	1005.8	25.1	7.1	36.8	14.7	/	219.2	119.4
Aug.	1007.0	26.4	8.1	37.1	15.9	/	149.8	188.2
Sept.	1010.1	22.0	7.8	34.7	11.0	/	268.6	209.7
Oct.	1015.5	15.8	9.0	31.2	1.6	/	166.1	156.5
Nov.	1018.5	11.1	9.0	25.4	− 2.4	/	124.7	68.0
Dec.	1018.7	6.4	7.5	22.5	− 4.2	/	153.1	64.2
Annual	1013.0	14.2	22.6	37.1	− 8.5	/	1980.2	209.7

Month	Mean evapor. (mm)	Number of days			Mean cloud-iness	Mean hours sun-shine	Wind	
		precip.	thunder-storms	fog			most freq. direction	mean speed (m/sec)
Jan.	34.4	21	1	2	/	75.4	W	4.0
Feb.	37.7	17	1	2	/	93.2	W	3.5
Mar.	65.2	15	1	3	/	147.3	W	3.4
Apr.	101.7	11	1	4	/	202.8	W	3.4
May	124.8	10	1	3	/	219.0	E	3.1
June	123.0	11	1	2	/	184.4	W	3.1
July	143.1	12	2	1	/	191.9	WSW	3.0
Aug.	171.6	8	4	2	/	244.6	E	2.9
Sept.	100.3	13	2	3	/	152.9	E	2.6
Oct.	75.7	10	1	6	/	159.7	ENE	2.4
Nov.	50.1	13	1	4	/	131.3	W	2.6
Dec.	34.8	17	1	4	/	82.9	W	3.5
Annual	1062.4	159	15	35	/	1885.3	W	3.1

TABLE XVI

CLIMATIC TABLE FOR NAGANO
Latitude 36°40′N, longitude 138°12′E, elevation 418.1 m

Month	Mean stat. press. (mbar)	Mean daily temp. (°C)	Mean daily temp. range (°C)	Temp. extremes (°C) highest	lowest	Mean vapor press. (mbar)	Mean precip. (mm)	Max. precip. (24 h)
Jan.	966.1	− 1.6	8.7	18.1	−17.0	4.5	60.7	32.9
Feb.	966.8	− 1.1	9.2	22.5	−16.4	4.5	49.4	36.0
Mar.	966.3	2.7	10.6	24.4	−14.6	5.5	54.4	33.3
Apr.	965.7	9.3	12.7	30.8	− 6.5	7.7	68.3	53.6
May	963.4	14.9	12.9	32.2	− 1.8	11.4	77.0	53.7
June	960.6	19.4	10.7	35.6	3.9	16.5	111.1	90.5
July	961.0	23.6	9.6	37.9	10.2	22.3	141.7	108.2
Aug.	962.4	24.6	10.4	38.6	10.7	23.1	111.9	112.9
Sept.	964.7	19.9	9.5	35.9	6.2	18.6	127.0	99.5
Oct.	968.4	13.2	10.1	32.2	− 1.8	12.0	91.8	83.1
Nov.	969.9	7.1	10.3	25.8	−11.4	8.0	49.6	40.1
Dec.	968.1	1.5	8.7	21.3	−15.2	5.6	58.3	58.2
Annual	965.3	11.1	26.2	38.6	−17.0	11.6	1001.2	112.9

Month	Mean evapor. (mm)	Number of days precip.	thunder-storms	gales	fog	Mean cloud-iness	Mean hours sun-shine	Wind most freq. direction	mean speed (m/sec)
Jan.	31.4	12	0	5	2	6.5	139.2	ENE	2.4
Feb.	39 3	8	0	6	1	6 4	147.2	ENE	2.8
Mar.	69.4	9	0	12	1	6.4	190.1	ENE	3.6
Apr.	112.1	9	0	15	1	6.1	213.4	ENE	4.0
May	139.7	10	1	12	1	6.6	228.6	WSW	3.5
June	142.3	12	2	8	1	7.7	189.2	WSW	3.2
July	157.4	13	4	5	1	7.5	199.0	WSW	2.7
Aug.	177.5	9	5	6	2	6.6	232.9	WSW	2.9
Sept.	111.4	11	1	5	4	7.4	156.8	WSW	3.0
Oct.	73.9	8	0	8	7	6.7	154.7	W	3.1
Nov.	46.7	7	0	7	7	5.8	151.3	ENE	2.6
Dec.	29.4	9	0	4	4	6.1	134.7	ENE	2.4
Annual	1130.4	117	15	93	32	6.7	2137.1	ENE	3.0

TABLE XVII

CLIMATIC TABLE FOR NAGASAKI
Latitude 32°44′N, longitude 129°52′E, elevation 26.9 m

Month	Mean stat. press. (mbar)	Mean daily temp. (°C)	Mean daily temp. range (°C)	Temp. extremes (°C)		Mean vapor press. (mbar)	Mean precip. (mm)	Max. precip. (24 h)
				highest	lowest			
Jan.	1018.5	6.4	7.0	21.3	− 3.2	6.5	69.7	82.7
Feb.	1017.0	7.6	7.5	22.6	− 3.2	6.8	83.2	86.0
Mar.	1015.4	10.2	7.8	24.4	− 2.0	8.5	116.4	133.1
Apr.	1012.4	14.7	8.2	27.7	2.6	11.8	182.0	345.4
May	1008.6	18.5	7.7	30.0	7.7	15.7	206.7	171.8
June	1005.1	21.9	6.4	34.3	12.0	21.0	293.8	385.4
July	1005.1	26.3	5 7	36 3	18.3	27.4	288.1	178.6
Aug.	1005.2	27.4	6.9	37.5	18.8	27.4	189.0	252.0
Sept.	1008.1	23.9	7.1	34.6	12.2	22.5	253.0	344.5
Oct.	1013.6	18.6	8.2	31.6	5.5	14.8	106.2	139.0
Nov.	1017.3	13.8	8.4	27.4	1.3	11.1	82.6	87.7
Dec.	1019.4	9.0	7.6	23.8	− 2.6	8.0	86.7	111.2
Annual	1012.1	16.5	21.0	37.5	− 3.2	15.1	1957.4	385.4

Month	Mean evapor. (mm)	Number of days				Mean cloud-iness	Mean hours sun-shine	Wind	
		precip.	thunder-storms	gales	fog			most freq. direction	mean speed (m/sec)
Jan.	41.0	12	0	8	1	7.0	114.4	x	x
Feb.	49.3	9	0	7	1	6.8	131.5	x	x
Mar.	75.5	11	1	9	0	6.5	176.5	x	x
Apr.	94.1	11	1	7	1	6.4	188.1	x	x
May	113.1	13	1	6	1	6.9	201.8	x	x
June	108.3	13	1	7	1	7.8	171.8	x	x
July	140.6	12	3	7	0	7.2	210.3	x	x
Aug.	160.1	9	3	3	−	5.9	252.1	x	x
Sept.	113.6	12	2	4	−	6.4	186.1	x	x
Oct.	94.3	6	1	3	0	5.2	199.7	x	x
Nov.	62.3	6	0	3	1	5.2	169.8	x	x
Dec.	43.8	8	1	5	1	6.3	127.7	x	x
Annual	1095.9	122	13	70	8	6.5	2129.8	x	x

TABLE XVIII

CLIMATIC TABLE FOR NAGOYA
Latitude 35°10′N, longitude 136°58′E, elevation 51.3 m

Month	Mean stat. press. (mbar)	Mean daily temp. (°C)	Mean daily temp. range (°C)	Temp. extremes (°C) highest	Temp. extremes (°C) lowest	Mean vapor press. (mbar)	Mean precip. (mm)	Max. precip. (24 h)
Jan.	1011.7	2.9	9.5	19.0	−10.3	5.6	49.3	53.0
Feb.	1011.6	3.6	10.3	19.8	− 9.5	5.6	64.0	101.0
Mar.	1010.6	7.1	11.3	25.8	− 6.8	7.0	100.1	95.8
Apr.	1009.4	12.7	11.9	29.6	− 2.1	10.3	136.5	104.6
May	1006.1	17.5	11.4	34.8	2.8	14.5	145.2	100.0
June	1002.7	21.5	9.6	34.8	8.2	19.9	203.5	183.5
July	1002.8	25.7	9.1	38.9	14.0	26.0	178.4	176.2
Aug	1003.6	26.6	9.8	39.9	14.4	27.1	155.4	171.6
Sept.	1006.1	22.7	9.3	36.2	9.5	22.3	211.9	240.1
Oct.	1010.3	16.5	10.3	30.2	1.5	14.8	159.7	166.7
Nov.	1013.0	10.9	11.0	27.2	− 2.7	10.0	85.5	92.4
Dec.	1013.1	5.6	9.8	20.9	− 7.2	6.9	56.7	54.2
Annual	1008.4	14.4	23.7	39.9	−10.3	14.2	1546.3	240.1

Month	Mean evapor. (mm)	Number of days precip.	Number of days thunder-storms	Number of days gales	Number of days fog	Mean cloud-iness	Mean hours sun-shine	Wind most freq. direction	Wind mean speed (m/sec)
Jan.	44.7	7	–	9	2	5.3	177.9	NW	3.7
Feb.	55.0	6	0	10	2	5.4	180.8	NW	4.0
Mar.	84.0	9	0	13	2	5.5	203.8	NW	4.5
Apr.	107.8	10	1	11	2	6.2	199.7	NNW	4.2
May	128.8	11	1	7	2	6.6	215.3	NW	3.7
June	118.7	13	1	4	2	7.8	173.1	S	3.2
July	142.9	13	4	3	1	7.3	202.0	SSE	3.1
Aug.	161.3	9	5	4	1	6.4	237.2	SSE	3.4
Sept.	103.6	14	3	4	3	7.1	166.1	N	3.2
Oct.	78.5	9	0	4	3	6.1	171.5	N	3.1
Nov.	56.6	7	0	5	3	5.1	175.4	N	3.3
Dec.	43.2	5	0	8	3	5.1	166.9	N	3.5
Annual	1125.2	114	15	82	26	6.2	2269.7	N	3.6

TABLE XIX

CLIMATIC TABLE FOR NAHA

Latitude 26°14′N, longitude 127°41′E, elevation 36.0 m

Month	Mean stat. press. (mbar)	Mean daily temp. (°C)	Mean daily temp. range (°C)	Temp. extremes (°C)		Mean vapor press. (mbar)	Mean precip. (mm)	Max. precip. (24 h)
				highest	lowest			
Jan.		16.1	5.4	25.8	6.7	13.5	120.7	94.5
Feb.		16.5	5.5	25.7	6.6	14.4	137.1	187.8
Mar.		17.9	5.6	26.9	7.2	16.1	168.4	180.2
Apr.		20.4	5.6	29.1	10.8	19.6	164.7	154.3
May		23.4	5.2	31.2	14.0	24.4	245.6	193.7
June		25.9	4.8	32.6	16.9	29.1	329.3	200.3
July		27.9	5.1	33.8	22.2	31.0	179.8	166.7
Aug.		27.4	5.0	32.6	22.3	30.7	295.7	271.7
Sept.		26.7	5.3	32.6	20.7	28.6	166.8	203.4
Oct.		24.0	5.3	30.4	14.8	23.1	153.6	372.1
Nov.		21.2	5.3	29.7	12.2	19.1	146.2	258.9
Dec.		18.1	5.1	26.4	7.2	15.2	114.4	86.2
Annual		22.1	11.8	33.8	6.6	22.1	2222.2	372.1

Month	Mean evapor. (mm)	Number of days			Mean cloud-iness	Mean hours sun-shine	Wind	
		precip.	thunder-storms	fog			most freq. direction	mean speed (m/sec)
Jan.	104.3	13	0	0	7.4	110.3	N	6.4
Feb.	97.1	13	1	–	7.6	106.2	N	6.1
Mar.	113.9	12	2	0	7.6	129.6	NNE	5.7
Apr.	124.6	12	2	1	7.5	152.6	S	5.7
May	140.0	14	3	1	8.0	169.0	SSW	5.1
June	144.1	13	3	0	8.1	195.4	SSW	5.4
July	190.3	9	3	0	6.6	279.1	S	5.0
Aug.	171.3	15	3	0	6.5	246.6	E	5.8
Sept.	161.3	11	1	0	5.8	224.2	E	5.6
Oct.	156.6	9	0	0	6.2	183.3	NE	6.0
Nov.	127.2	10	0	–	6.6	146.6	NE	6.3
Dec.	117.7	10	1	0	7.3	123.4	NE	6.0
Annual	1648.4	141	19	2	7.1	2066.2	NE	5.8

TABLE XX

CLIMATIC TABLE FOR NAZE
Latitude 28°23′N, longitude 129°30′E, elevation 2.7 m

Month	Mean stat. press. (mbar)	Mean daily temp. (°C)	Mean daily temp. range (°C)	Temp. extremes (°C)		Mean vapor press. (mbar)	Mean precip. (mm)	Max. precip. (24 h)
				highest	lowest			
Jan.	1020.1	14.3	6.2	26.4	4.6	x	162.8	125.9
Feb.	1018.4	14.7	6.2	27.7	3.1	x	184.4	135.7
Mar.	1017.1	16.5	6.7	29.1	4.7	x	219.7	154.6
Apr.	1014.5	19.3	7.0	32.1	6.6	x	221.0	210.6
May	1011.1	22.3	6.8	33.7	9.4	x	362.1	547.1
June	1008.3	25.2	6.5	35.5	13.9	x	442.5	365.4
July	1008.6	28.1	7.2	37.3	18.8	x	231.3	416.4
Aug.	1007.2	27.7	7.0	36.9	19.6	x	281.4	407.0
Sept.	1009.7	26.4	7.2	34.9	15.3	x	296.8	397.9
Oct.	1014.2	23.0	6.8	32.7	11.2	x	247.4	253.6
Nov.	1017.7	19.8	6.6	31.0	8.2	x	223.6	421.5
Dec.	1020.4	16.4	6.4	27.6	6.1	x	160.2	252.5
Annual	1014.0	21.1	13.8	37.3	3.1	x	3033.1	547.1

Month	Mean evapor. (mm)	Number of days				Mean cloud-iness	Mean hours sun-shine	Wind	
		precip.	thun-der-storms	gales	fog			most freq. direction	mean speed (m/sec)
Jan.	67.8	17	0	7	0	8.2	79.7	N	4.7
Feb.	67.6	16	1	5	0	8.3	78.1	N	4.5
Mar.	92.1	15	2	5	0	8.0	106.1	N	4.4
Apr.	101.6	15	1	4	1	7.9	124.8	S	4.0
May	110.3	17	2	2	1	8.2	137.0	S	3.3
June	114.8	16	2	2	1	8.4	135.3	S	3.2
July	165.9	11	2	1	0	6.7	235.2	S	3.0
Aug.	147.7	14	3	2	0	6.5	212.8	S	3.3
Sept.	131.8	16	2	3	0	6.5	186.4	S	3.4
Oct.	114.6	12	1	5	0	6.9	141.1	N	3.9
Nov.	87.5	11	1	5	0	7.5	102.8	N	4.4
Dec.	76.2	12	0	6	0	7.8	92.9	N	4.7
Annual	1277.9	172	16	48	3	7.6	1632.0	S	3.9

TABLE XXI

CLIMATIC TABLE FOR NEMURO
Latitude 43°20′N, longitude 145°35′E, elevation 27.5 m

Month	Mean stat. press. (mbar)	Mean daily temp. (°C)	Mean daily temp. range (°C)	Temp. extremes (°C)		Mean vapor press. (mbar)	Mean precip. (mm)	Max. precip. (24 h)
				highest	lowest			
Jan.	1007.7	− 4.8	6.7	10.3	−22.7	3.2	48.9	82.7
Feb.	1009.9	− 5.6	7.3	7.8	−22.9	3.1	40.3	52.9
Mar.	1010.3	− 2.2	6.2	12.6	−19.2	4.1	77.3	119.3
Apr.	1009.0	2.8	7.1	21.7	−12.9	6.0	77.1	62.2
May	1008.1	6.8	7.5	25.4	− 3.9	8.3	99.2	91.4
June	1007.5	10.0	6.9	28.8	0.6	11.1	97.2	116.2
July	1007.0	14.3	6.6	32.1	2.3	15.3	104.0	110.2
Aug.	1007.6	17.5	6.4	33.0	6.3	18.4	106.4	124.8
Sept.	1009.8	15.5	6.0	29.0	3.2	15.3	152.0	135.2
Oct.	1013.7	10.8	6.7	22.5	− 3.3	10.4	124.0	110.4
Nov.	1012.7	4.7	6.8	19.2	− 8.6	6.4	91.9	72.9
Dec.	1009.3	− 1.3	6.3	13.4	−15.1	4.0	63.0	112.0
Annual	1009.4	5.7	23.1	33.0	−22.9	8.8	1081.4	135.2

Month	Mean evapor. (mm)	Number of days				Mean cloud-iness	Mean hours sun-shine	Wind	
		precip.	thunder-storms	gales	fog			most freq. direction	mean speed (m/sec)
Jan.	x	10	–	18	1	5.5	153.5	NW	6.0
Feb.	x	8	–	11	2	5.5	168.4	NW	5.0
Mar.	x	9	0	14	4	6.2	192.1	N	5.6
Apr.	x	9	0	15	9	6.6	193.7	S	5.8
May	89.8	10	0	14	14	7.5	182.0	S	5.7
June	87.2	9	1	6	20	8.0	151.7	SE	4.6
July	92.4	8	1	4	23	8.7	131.9	ESE	4.2
Aug.	94.6	8	1	4	20	8.4	146.2	SSE	4.3
Sept.	80.1	10	1	9	9	7.2	158.1	SSE	4.8
Oct.	69.7	10	1	13	3	6.0	173.9	S	5.4
Nov.	x	8	0	17	2	5.7	151.2	W	5.8
Dec.	x	8	0	18	1	5.3	150.0	W	6.0
Annual		106	5	143	108	6.7	1952.5	S	5.3

TABLE XXII

CLIMATIC TABLE FOR NIIGATA
Latitude 37°55′N, longitude 139°03′E, elevation 2.0 m

Month	Mean stat. press. (mbar)	Mean daily temp. (°C)	Mean daily temp. range (°C)	Temp. extremes (°C)		Mean vapor press. (mbar)	Mean precip. (mm)	Max. precip. (24 h)
				highest	lowest			
Jan.	1016.7	1.7	5.4	15.2	−11.7	5.3	194.3	68.2
Feb.	1017.7	1.8	5.9	18.8	−13.0	5.3	125.6	43.7
Mar.	1016.9	4.8	6.8	25.1	− 6.4	6.3	120.5	51.0
Apr.	1015.4	10.2	8.9	28.0	− 2.5	8.9	104.1	73.0
May	1012.0	15.3	8.4	31.3	2.0	12.9	94.5	53.7
June	1008.5	19.9	7.1	35.0	6.7	18.4	126.5	111.9
July	1008.1	24.1	6.7	38.5	11.4	24.5	193.0	140.7
Aug.	1009.3	25.8	7.7	39.1	14.5	26.1	106.5	132.7
Sept.	1012.3	21.4	7.4	36.2	7.9	20.2	176.5	116.6
Oct.	1017.1	15.5	7.2	33.3	3.0	13.5	165.4	81.7
Nov.	1018.9	9.8	7.0	26.1	− 1.3	9.3	170.5	78.3
Dec.	1018.1	4.7	5.8	23.6	− 8.0	6.7	263.7	77.0
Annual	1014.2	12.9	24.1	39.1	−13.0	13.1	1840.9	140.7

Month	Mean evapor. (mm)	Number of days				Mean cloud-iness	Mean hours sun-shine	Wind	
		precip.	thunder-storms	gales	fog			most freq. direction	mean speed (m/sec)
Jan.	42.3	22	1	20	1	8.8	70.3	NW	6.1
Feb.	44.3	19	1	15	1	8.6	82.0	SSW	5.3
Mar.	64.6	17	1	14	1	7.9	139.9	SSW	4.9
Apr.	107.2	12	1	10	1	6.7	201.3	SSW	4.5
May	130.1	10	1	6	1	7.0	219.8	S	3.9
June	139.1	11	1	3	1	7.6	202.2	NNE	3.7
July	158.7	13	2	2	2	7.4	206.5	SW	3.2
Aug.	188.7	11	3	3	2	6.1	254.4	NNE	3.4
Sept.	119.0	13	2	5	1	7.5	169.0	S	3.5
Oct.	81.9	14	2	8	0	7.2	144.9	S	3.7
Nov.	53.1	18	2	11	1	7.6	109.9	SSW	4.3
Dec.	36.0	24	2	18	1	8.7	58.6	SSW	5.8
Annual	1165.1	181	20	113	12	7.6	1858.7	S	4.4

TABLE XXIII

CLIMATIC TABLE FOR OSAKA
Latitude 34°39′N, longitude 135°32′E, elevation 6.7 m

Month	Mean sta. press. (mbar)	Mean daily temp. (°C)	Mean daily temp. range (°C)	Temp. extremes (°C) highest	lowest	Mean vapor press. (mbar)	Mean precip. (mm)	Max. precip. (24 h)
Jan.	1019.1	4.5	8.4	18.0	− 7.5	5.8	43.2	58.4
Feb.	1018.3	4.9	8.7	23.7	− 6.5	6.0	57.6	50.6
Mar.	1017.1	8.0	9.5	23.8	− 5.2	7.4	95.7	73.4
Apr.	1015.1	13.6	10.7	28.8	− 2.6	10.5	126.8	101.7
May	1011.5	18.3	10.0	31.2	3.5	14.4	122.1	132.1
June	1007.9	22.3	8.5	34.3	8.9	19.7	192.9	250.7
July	1007.8	26.6	7.9	36.7	14.8	26.0	176.9	134.2
Aug.	1008.6	27.8	8.9	38.2	13.6	26.6	118.3	174.7
Sept.	1011.3	23.7	8.8	35.1	10.4	21.8	170.9	117.9
Oct.	1016.2	17.4	9.6	32.5	3.0	14.8	122.1	133.0
Nov.	1019.3	11.9	9.9	26.1	− 2.2	10.4	80.8	103.4
Dec.	1019.8	7.0	8.7	22.2	− 4.5	7.2	51.7	55.0
Annual	1014.3	15.5	23.3	38.2	− 7.5	14.2	1359.0	250.7

Month	Mean evapor. (mm)	Number of days precip.	thunder- storms	gales	fog	Mean cloud- iness	Mean hours sun- shine	Wind most freq. direction	mean speed (m/sec)
Jan.	50.7	7	0	3	9	5.5	150.2	W	3.4
Feb.	51.1	7	0	7	8	6.1	143.8	NNE	3.2
Mar.	76.6	11	0	4	6	6.2	175.4	N	3.1
Apr.	108.8	10	1	5	3	6.1	195.7	NNE	3.1
May	136.2	11	1	4	3	6.6	208.8	NNE	2.9
June	135.5	12	1	3	3	7.6	180.5	W	2.9
July	171.7	10	3	2	2	6.8	212.9	W	2.8
Aug.	188.2	7	4	3	1	5.9	241.4	W	3.0
Sept.	124.9	12	2	2	2	6.7	175.1	NNE	2.7
Oct.	84.6	9	0	2	5	6.2	166.2	NNE	2.6
Nov.	56.2	7	0	1	11	5.4	155.5	NNE	2.3
Dec.	48.1	5	0	4	13	5.3	146.1	W	2.9
Annual	1232.6	107	12	40	66	6.2	2151.4	NNE	2.9

TABLE XXIV

CLIMATIC TABLE FOR SAPPORO
Latitude 43°03′N, longitude 141°20′E, elevation 16.9 m

Month	Mean stat. press. (mbar)	Mean daily temp. (°C)	Mean daily temp. range (°C)	Temp. extremes (°C) highest	Temp. extremes (°C) lowest	Mean vapor press. (mbar)	Mean precip. (mm)	Max. precip. (24 h)
Jan.	1010.6	− 5.5	8.8	11.2	−23.9	3.2	111.2	114.8
Feb.	1012.2	− 4.7	9.1	10.8	−22.5	3.3	82.5	62.3
Mar.	1012.2	− 1.0	8.5	16.8	−17.9	4.2	67.4	46.7
Apr.	1010.2	5.7	10.1	25.2	− 9.7	6.2	66.1	104.2
May	1008.2	11.3	11.0	31.1	− 2.2	9.3	59.2	84.7
June	1006.8	15.5	10.1	31.9	2.5	13.6	67.4	119.9
July	1006.3	20.0	9.0	35.8	7.2	19.0	100.0	123.5
Aug.	1007.5	21.7	9.1	34.5	8.5	21.0	107.4	146.6
Sept.	1010.2	16.8	10.2	32.1	1.7	15.2	145.4	139.6
Oct.	1014.7	10.4	11.0	25.1	− 4.2	9.8	113.1	103.9
Nov.	1015.2	3.6	8.7	21.3	−12.0	6.0	111.8	89.4
Dec.	1012.4	− 2.6	8.1	14.6	−20.2	3.9	104.3	67.3
Annual	1010.5	7.6	27.2	35.8	−23.9	9.6	1135.7	146.6

Month	Mean evapor. (mm)	Number of days precip.	Number of days thunderstorms	Number of days gales	Number of days fog	Mean cloudiness	Mean hours sunshine	Wind most freq. direction	Wind mean speed (m/sec)
Jan.	x	17	0	4	0	7.3	98.6	NW	2.8
Feb.	x	14	0	4	0	7.5	112.1	NW	3.0
Mar.	x	12	–	7	0	7.2	157.9	NNW	3.6
Apr.	x	10	0	10	1	6.5	198.6	SE	4.2
May	105.9	8	0	11	2	6.8	212.6	NNW	4.4
June	120.4	9	1	5	4	7.1	204.5	SE	3.8
July	124.1	9	1	4	4	7.5	189.7	SE	3.4
Aug.	122.9	10	1	2	2	7.0	200.8	SE	3.1
Sept.	85.2	11	1	4	1	6.7	173.0	SE	3.0
Oct.	55.2	12	1	3	1	6.0	165.6	SE	2.9
Nov.	x	13	0	3	1	6.9	111.5	NW	2.9
Dec.	x	15	0	5	1	7.3	91.3	NW	2.9
Annual		139	6	59	17	7.0	1916.1	SE	3.4

TABLE XXV

CLIMATIC TABLE FOR SENDAI
Latitude 38°16′N, longitude 140°54′E, elevation 38.4 m

Month	Mean stat. press. (mbar)	Mean daily temp. (°C)	Mean daily temp. range (°C)	Temp. extremes (°C)		Mean vapor press. (mbar)	Mean precip. (mm)	Max. precip. (24 h)
				highest	lowest			
Jan.	1010.3	0.1	8.3	13.7	−19.8	4.5	37.0	43.0
Feb.	1011.5	0.6	8.7	14.1	−20.2	4.5	43.7	48.8
Mar.	1011.0	3.5	9.4	21.0	−15.4	5.4	62.4	57.7
Apr.	1010.2	9.0	10.5	26.1	− 6.0	7.9	95.4	89.4
May	1007.4	13.9	9.9	30.9	− 1.2	11.7	100.1	75.0
June	1004.8	17.8	7.6	33.7	5.4	16.8	154.9	130.9
July	1004.6	22.0	6.6	35.3	10.4	22.8	167.2	143.8
Aug.	1005.8	23.8	7.2	36.4	11.5	24.9	136.4	186.8
Sept.	1008.5	19.8	8.0	35.4	4.1	19.1	190.5	152.2
Oct.	1012.9	13.8	9.4	28.9	− 1.0	12.4	133.4	106.5
Nov.	1014.3	8.2	9.8	23.3	− 5.4	8.1	61.1	95.2
Dec.	1012.1	2.9	8.4	21.4	−12.9	5.6	49.6	66.4
Annual	1009.4	11.3	23.7	36.4	−20.2	12.0	1231.5	186.8

Month	Mean evapor. (mm)	Number of days				Mean cloud-iness	Mean hours sun-shine	Wind	
		precip.	thun-der-storms	gales	fog			most freq. direction	mean speed (m/sec)
Jan.	42.6	7	–	8	0	5.7	150.6	WNW	3.2
Feb.	46.8	6	–	6	0	5.8	154.5	WNW	3.3
Mar.	74.1	8	0	9	1	6.1	192.1	NW	3.4
Apr.	103.2	9	0	8	2	6.3	202.8	WNW	3.5
May	115.7	10	1	4	4	6.9	205.8	SSE	2.9
June	98.6	13	1	1	5	8.0	153.2	SE	2.4
July	101.2	14	2	0	9	8.2	134.1	SSE	1.9
Aug.	116.6	12	3	1	6	7.6	162.6	SSE	2.1
Sept.	84.8	11	1	2	2	7.6	129.1	NNW	2.2
Oct.	67.6	9	0	2	2	6.4	149.3	NW	2.5
Nov.	51.1	6	0	4	1	5.7	141.8	NW	2.7
Dec.	37.8	8	0	4	1	5.8	130.9	WNW	2.9
Annual	940.3	110	8	51	33	6.7	1906.8	NW	2.8

TABLE XXVI

CLIMATIC TABLE FOR TOKUSHIMA
Latitude 34°04′N, longitude 134°35′E, elevation 1.3 m

Month	Mean stat. press. (mbar)	Mean daily temp. (°C)	Mean daily temp. range (°C)	Temp. extremes (°C)		Mean vapor press. (mbar)	Mean precip. (mm)	Max. precip. (24 h)
				highest	lowest			
Jan.	1019.3	4.9	8.0	21.5	− 5.4	5.9	40.1	85.9
Feb.	1018.9	5.2	8.1	21.7	− 6.0	6.1	60.9	74.6
Mar.	1017.6	8.1	9.0	24.7	− 3.6	7.8	94.5	86.9
Apr.	1015.4	13.2	9.7	27.2	− 0.7	11.1	132.5	137.4
May	1011.8	17.7	9.2	30.4	4.6	15.2	126.3	102.0
June	1008.2	21.6	7.6	33.8	9.7	20.7	206.5	219.3
July	1008.1	25.7	6.9	35.9	15.3	27.3	196.6	273.5
Aug.	1008.9	26.5	7.6	37.0	16.6	27.8	168.7	471.5
Sept.	1011.7	23.3	7.6	35.1	11.9	22.6	257.2	306.0
Oct.	1016.7	17.5	7.9	30.6	4.5	15.2	193.7	463.4
Nov.	1019.9	12.6	8.3	25.3	− 1.3	10.7	93.6	132.5
Dec.	1020.6	7.7	8.3	22.2	− 4.2	7.4	54.2	106.5
Annual	1014.8	15.3	21.6	37.0	− 6.0	14.8	1624.7	471.5

Month	Mean evapor. (mm)	Number of days				Mean cloudiness	Mean hours sunshine	Wind	
		precip.	thunderstorms	gales	fog			most freq. direction	mean speed (m/sec)
Jan.	64.0	6	0	12	0	5.6	166.5	W	4.4
Feb.	65.5	7	0	11	0	6.1	157.0	W	4.3
Mar.	91.2	10	0	11	1	6.2	185.5	W	4.4
Apr.	110.8	10	0	12	1	6.3	197.9	NW	4.3
May	129.3	11	0	9	1	6.9	207.5	SE	3.9
June	123.7	13	1	5	2	7.9	173.9	SE	3.2
July	149.2	11	3	5	2	6.9	216.5	SE	3.1
Aug.	159.3	8	3	5	0	5.9	245.8	ESE	3.4
Sept.	113.5	13	2	5	0	6.9	173.5	W	3.3
Oct.	93.3	8	0	4	0	6.4	166.9	W	3.3
Nov.	70.5	6	0	5	0	5.6	162.1	W	3.5
Dec.	62.5	5	0	7	0	5.5	159.0	W	4.1
Annual	1232.8	107	10	91	8	6.4	2212.0	W	3.8

TABLE XXVII

CLIMATIC TABLE FOR TOKYO
Latitude 35°41′N, longitude 139°46′E, elevation 4.1 m

Month	Mean stat. press. (mbar)	Mean daily temp. (°C)	Mean daily temp. range (°C)	Temp. extremes (°C)		Mean vapor press. (mbar)	Mean precip. (mm)	Max. precip. (24 h)
				highest	lowest			
Jan.	1014.8	3.7	10.1	21.3	− 9.2	4.9	47.9	48.4
Feb.	1015.6	4.3	9.5	24.9	− 7.9	5.1	72.9	90.8
Mar.	1014.9	7.6	9.5	25.2	− 5.6	6.8	101.4	87.1
Apr.	1014.3	13.1	9.5	27.2	− 3.1	10.2	135.1	81.1
May	1011.5	17.6	8.8	31.4	2.2	14.5	131.0	120.8
June	1008.3	21.1	7.4	34.7	8.5	19.4	182.3	278.3
July	1008.2	25.1	7.2	37.0	13.0	25.2	146.3	151.3
Aug.	1009.3	26.4	7.5	38.4	15.4	26.8	147.4	171.5
Sept.	1012.0	22.8	7.4	36.4	10.5	21.7	216.7	392.5
Oct.	1016.2	16.7	7.4	32.3	− 0.5	14.5	220.3	163.6
Nov.	1018.0	11.3	8.9	27.3	− 3.1	9.6	101.2	168.5
Dec.	1016.6	6.1	9.8	22.7	− 6.8	6.2	60.9	85.4
Annual	1013.3	14.7	22.7	38.4	− 9.2	13.8	1563.4	392.5

Month	Mean evapor. (mm)	Number of days				Mean cloud-iness	Mean hours sun-shine	Wind	
		precip.	thun-der-storms	gales	fog			most freq. direction	mean speed (m/sec)
Jan.	51.3	6	0	6	3	4.1	186.2	NNW	3.5
Feb.	58.5	7	0	8	3	5.2	165.7	NNW	3.8
Mar.	82.4	10	0	10	2	6.1	175.9	NNW	4.3
Apr.	104.1	11	1	12	2	6.6	179.9	N	4.3
May	120.7	12	1	9	3	7.1	192.8	S	3.9
June	108.9	12	1	4	2	8.2	149.3	S	3.5
July	131.9	11	2	5	2	7.7	180.6	S	3.6
Aug.	141.5	10	3	5	3	6.8	203.6	S	3.7
Sept.	94.2	13	2	3	3	7.7	136.1	N	3.5
Oct.	68.9	12	0	4	3	7.0	135.7	NNW	3.7
Nov.	53.6	8	0	4	4	5.9	144.4	N	3.2
Dec.	46.8	5	0	5	6	4.4	168.5	NNW	3.0
Annual	1062.8	115	10	75	37	6.4	2018.7	NNW	3.7

TABLE XXVIII

CLIMATIC TABLE FOR WAKKANAI
Latitude 45°25′N, longitude 141°41′E, elevation 1.8 m

Month	Mean stat. press. (mbar)	Mean daily temp. (°C)	Mean daily temp. range (°C)	Temp. extremes (°C) highest	Temp. extremes (°C) lowest	Mean vapor press. (mbar)	mean precip. (mm)	Max. precip. (24 h)
Jan.	1011.8	− 5.9	4.3	6.3	−19.4	3.0	93.7	25.7
Feb.	1013.3	− 5.6	5.0	5.4	−17.3	3.1	62.2	36.0
Mar.	1013.3	− 1.8	5.3	12.3	−16.1	4.0	64.6	101.0
Apr.	1010.7	4.0	5.8	20.2	− 8.0	6.2	62.8	49.3
May	1009.6	8.4	6.3	26.9	− 2.2	8.6	77.5	41.8
June	1009.0	12.2	6.2	25.1	2.1	12.0	70.2	39.5
July	1008.4	16.7	5.6	29.5	6.8	16.7	111.8	128.2
Aug.	1009.6	19.6	5.6	31.3	8.9	19.3	105.4	138.1
Sept.	1011.3	16.6	6.2	29.0	3.5	14.4	151.7	139.9
Oct.	1015.4	10.7	6.4	23.5	− 4.4	9.1	129.1	84.6
Nov.	1015.4	3.0	5.2	17.0	− 9.8	5.3	119.5	72.6
Dec.	1012.5	− 2.9	4.4	11.5	−16.0	3.6	112.2	52.2
Annual	1011.7	6.2	25.5	31.3	−19.4	8.8	1160.5	139.9

Month	Mean evapor. (mm)	Number of days precip.	Number of days thunder-storms	Number of days gales	Number of days fog	Mean cloud-iness	Mean hours sun-shine	Wind most freq. direction	Wind mean speed (m/sec)
Jan.	x	22	0	12	–	9.2	42.4	NNW	5.6
Feb.	x	15	–	11	–	8.5	78.7	E	5.6
Mar.	x	14	0	13	0	7.4	146.0	S	5.3
Apr.	x	10	–	14	1	6.5	183.9	SSW	5.6
May	85.5	9	0	16	3	7.1	202.6	SSW	5.6
June	89.8	9	1	8	6	7.6	176.9	E	4.5
July	94.0	9	1	8	7	7.9	163.8	E	4.2
Aug.	98.8	9	1	7	3	7.3	179.8	E	4.2
Sept.	88.7	10	3	8	0	6.2	188.4	S	4.5
Oct.	64.8	13	2	10	0	6.2	156.8	SSW	4.7
Nov.	x	17	0	12	–	8.1	72.8	W	5.3
Dec.	x	21	0	14	0	9.1	37.6	WNW	5.7
Annual	x	158	7	132	20	7.6	1629.8	SSW	5.1

TABLE XXIX

CLIMATIC TABLE FOR YAMAGATA
Latitude 38°15′N, longitude 140°21′E, elevation 150.6 m

Month	Mean stat. press. (mbar)	Mean daily temp. (°C)	Mean daily temp. range (°C)	Temp. extremes (°C)		Mean vapor press. (mbar)	Mean precip. (mm)	Max. precip. (24 h)
				highest	lowest			
Jan.	997.4	− 1.6	7.8	18.1	−20.0	4.6	101.2	57.1
Feb.	998.6	− 1.1	8.3	16.8	−19.0	4.6	78.1	54.2
Mar.	998.0	2.1	9.4	23.7	−15.5	5.4	75.4	49.2
Apr.	997.2	8.7	12.3	33.3	− 7.3	7.6	78.4	89.1
May	994.4	14.7	13.1	32.3	− 1.8	11.2	66.0	98.3
June	991.6	19.1	11.0	35.6	3.0	16.3	99.3	70.0
July	991.5	23.2	9.7	40.8	7.2	22.4	165.3	101.2
Aug.	992.8	24.4	10.3	37.3	8.4	23.6	136.4	217.6
Sept.	995.6	19.4	10.1	35.9	3.0	18.1	127.2	131.1
Oct.	1000.0	12.7	10.5	32.3	− 2.4	11.9	107.1	103.6
Nov.	1001.5	6.7	9.9	26.9	− 7.2	7.7	81.6	46.0
Dec.	999.3	1.4	7.4	19.9	−15.0	5.7	118.8	111.1
Annual	996.5	10.8	26.0	40.8	−20.0	11.6	1234.7	217.6

Month	Mean evapor. (mm)	Number of days				Mean cloud-iness	Mean hours sun-shine	Wind	
		precip.	thun-der-storms	gales	fog			most freq. direction	mean speed (m/sec)
Jan.	28.4	17	0	1	4	8.0	88.5	SSW	1.7
Feb.	37.2	12	0	0	3	7.8	102.9	SSW	1.9
Mar.	61.2	11	0	1	2	7.2	157.1	SSW	2.2
Apr.	100.3	9	0	3	2	6.6	195.0	N	2.5
May	129.3	8	1	2	2	6.8	214.8	N	2.2
June	129.0	12	1	0	3	7.6	182.0	NNE	1.9
July	135.9	13	3	0	5	7.7	173.8	N	1.6
Aug.	147.7	10	4	1	6	6.9	201.3	N	1.8
Sept.	93.4	11	1	1	7	7.5	145.6	NNE	1.6
Oct.	59.4	10	1	0	8	7.0	134.0	SSW	1.6
Nov.	36.7	11	0	1	6	7.0	111.0	SSW	1.7
Dec.	25.6	16	0	1	5	8.0	77.7	SSW	1.8
Annual	984.1	139	11	11	51	7.3	1783.8	N	1.9

TABLE XXX

CLIMATIC TABLE FOR YOKOHAMA
Latitude 35°26′N, longitude 139°39′E, elevation 34.5 m

Month	Mean stat. press. (mbar)	Mean daily temp. (°C)	Mean daily temp. range (°C)	Temp. extremes (°C) highest	lowest	Mean vapor press. (mbar)	Mean precip. (mm)	Max. precip. (24 h)
Jan.	1010.8	4.1	9.0	20.8	− 8.2	5.2	52.4	53.7
Feb.	1011.5	4.4	8.4	24.4	− 6.8	5.5	74.6	86.1
Mar	1010 8	7.5	8.8	23.2	− 4.6	7.2	110.7	95.8
Apr.	1010.5	12.7	8.6	28.7	− 0.5	10.8	148.6	93.4
May	1007.6	17.1	8.0	29.8	3.6	14.8	145.1	138.1
June	1004.6	20.5	6.7	33.0	9.2	19.6	194.8	268.3
July	1004.5	24.5	6.6	35.3	13.5	25.4	151.2	156.3
Aug.	1005.6	25.8	6.9	36.5	15.5	27.0	151.5	151.2
Sept.	1008.1	22.4	6.7	36.2	11.2	22.2	244.4	287.2
Oct.	1012.2	16.5	7.0	30.6	2.2	14.5	218.7	194.5
Nov.	1014.0	11.5	8.0	25.2	− 1.4	10.0	107.1	135.0
Dec.	1012.5	6.7	9.0	22.6	− 5.6	6.6	65.7	73.3
Annual	1009.4	14.5	21.7	36.5	− 8.2	14.1	1664.8	287.2

Month	Mean evapor. (mm)	Number of days precip.	thunder-storms	gales	fog	Mean cloud-iness	Mean hours sun-shine	Wind most freq. direction	mean speed (m/sec)
Jan.	62.9	6	0	15	3	4.2	185.0	N	5.0
Feb.	62.7	7	0	15	2	5.2	162.0	N	5.4
Mar.	82.6	10	0	16	3	6.2	175.9	N	5.6
Apr.	101.9	11	1	16	3	6.6	181.5	N	5.5
May	119.0	12	1	14	4	7.0	194.5	N	4.9
June	110.4	12	1	9	3	8.0	156.1	NNW	4.5
July	143.7	10	2	8	3	7.3	198.3	SSW	4.5
Aug.	152.5	10	3	7	2	6.3	227.4	S	4.6
Sept.	103.9	14	1	8	3	7.4	150.2	NNW	4.7
Oct.	76.5	12	0	12	3	6.9	141.1	NNW	5.3
Nov.	58.7	8	0	11	4	5.8	144.5	NNW	4.9
Dec.	54.8	5	0	12	5	4.4	170.2	NNW	4.6
Annual	1129.6	115	10	141	38	6.3	2086.8	NNW	5.0

Chapter 3

Climate of the Philippines

J. F. FLORES and V. F. BALAGOT

Introduction

Data used in this discussion of the climate of the Philippines were obtained from the records available at the Philippines Weather Bureau and the data found in cited publications. Available records date as far back as 1865 for some stations. Unfortunately, however, many records were lost or destroyed during World War II. Because of this loss, it was not possible to conform strictly with the terms of the World Meteorological Organization's convention for the use of normals for 1931–1960. Furthermore, in line with the space available for this chapter, the treatment of the climate of the Philippines in this work is very selective.

First a presentation of the different climatic controls that affect the Philippines will be made. This will be followed by a description of the behaviour of the climatic elements. Finally, a classification of the climatic types of the Philippines will provide a broad understanding of the climate of this region.

Climatic controls

Factors in the general atmospheric circulation, called climatic controls, acting with various intensities and in different combinations, produce the observed changes in the climatic elements. Aside from the geographic and topographic setting, the most important climatic controls affecting the Philippines are semi-permanent cyclones and anti-cyclones, air streams (air masses), ocean currents, linear systems, tropical cyclones and thunderstorms. The effects of each of these climatic controls are discussed below.

Geography and topography

The climate of the Philippines can be better grasped by a knowledge of its geography and topography. The Philippines lie just off the southeastern portion of the Asian continent in an almost north to south orientation. They extend from about 4.7°N to 21.5°N and 117°E to 127°E in their longest and broadest dimensions. They are made up of more than 7,000 islands with a total area of about 300,000 km². These are grouped into 3 regions as follows: Luzon Region (composed of Luzon Island and small islands in its vicinity); Visayas Region (composed of many islands near the center); and Mindanao Region (composed of Mindanao Island and small islands in its vicinity). Only 2 islands, Luzon and Mindanao, have areas of more than 80,000 km²; 9 islands have areas between 2,500 and 15,000 km²; 20 islands have areas between 250 and 2,500 km²; about 75 islands have areas between 2.5 and 250 km²; and the rest of the islands have areas less than 2.5 km². The Philippines is completely surrounded by large bodies of water. They are bounded on the north by the Luzon Strait, on the east by the Pacific Ocean, on the south by the Celebes Sea, and on the west by the South China Sea.

Many of the larger islands have narrow coastal plains which are generally less than 15 km wide, and interiors of highland plains and mountain ranges which are generally oriented north to south, some of which cover almost the entire length of the islands. Most of the mountain ranges reach heights of more than 500 m, with large areas above 1,000 m and a number of isolated portions above 2,000 m. Luzon, the largest island, has a number of mountain ranges with heights of more than 500 m which cover almost one-half of the entire island area. The most prominent of the mountain ranges of Luzon are the Sierra Madre Ranges along the eastern coast of northern and central Luzon, the Ilocos Ranges along the western coast of northern Luzon, the Cordillera Central Ranges between the Sierra Madre Ranges and the Ilocos Ranges, and the Zambales Ranges along the western coast of central Luzon. Many of the bigger islands of the Visayas have mountain ranges extending almost throughout their lengths with elevations of more than 500 m. Similarly, Mindanao is also covered with extensive mountain ranges along the eastern coast, in the middle, and along the western coast with elevations of more than 500 m. Like Luzon, these mountains cover about one-half of the total area of Mindanao. Fig.1 is a geographical and topographical map of the Philippines.

Semi-permanent cyclones and anticyclones

The climate of the Philippines is controlled to a great extent by the location and intensity of nearby semi-permanent cyclones and anticyclones. Global charts of mean sea level pressure distribution for January (TREWARTHA, 1954; HAURWITZ and AUSTIN, 1944) show a large anticyclone centered over Siberia with central pressure of more than 1035 mbar, a small anticyclone over the northeastern Pacific Ocean off the North American coast with central pressure greater than 1,020 mbar, and a trough of low pressure through New Guinea and northern Australia. In July, charts show a cyclone over Asia centered near the boundary of Pakistan and Afghanistan with central pressure of less than 1,000 mbar and an anticyclone over the Indian Ocean with central pressure greater than 1,024 mbar. It will be shown below that these semi-permanent pressure features produce air streams and ocean currents which greatly affect the climate of the Philippines.

Air streams (air masses)

The principal air streams affecting the Philippines are the Northeast Monsoon, the Southwest Monsoon, the North Pacific Trades, the Temperate Zone Westerlies, and the South Pacific Trades. Of these five air streams, the first three (Northeast Monsoon, Southwest Monsoon, and North Pacific Trades) are more important and have greater effects on the climate of the Philippines than the Temperate Zone Westerlies and South Pacific Trades. Since these air streams affect the behaviour of the climatic elements to a great extent, their characteristics are described below.

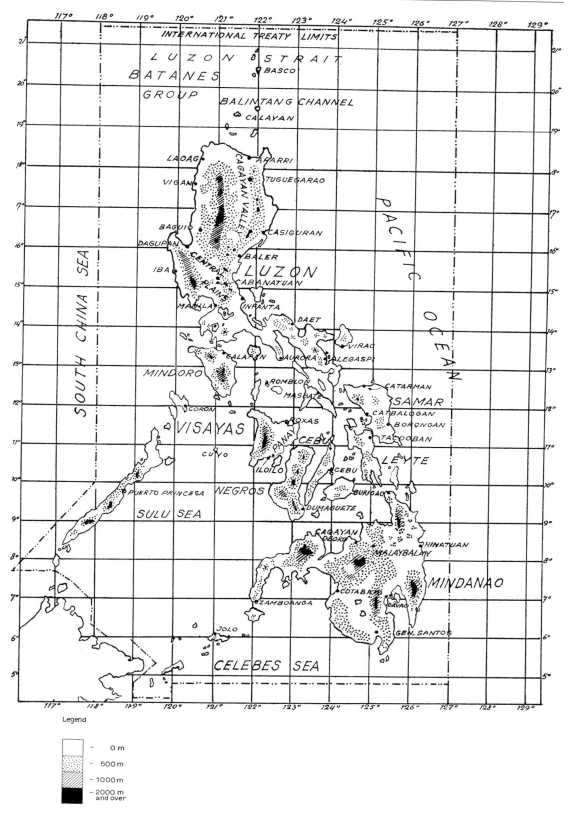

Legend

- 0 m
- 500 m
- 1000 m
- 2000 m and over

Fig.1. Geographical and topographical chart of the Philippines.

161

The Northeast Monsoon

This air stream is given other names by some authors. Northers is sometimes used (DEP-PERMAN, 1934) and so is Winter Monsoon (RAMAGE, 1959). This air stream originates in the cold, intense Asiatic winter anticyclone. Generally it follows a path across Japan or the Ryukyu Islands towards the northwestern Pacific Ocean. It finally reaches the Philippines generally as a northeasterly air stream, but sometimes as a northerly or easterly stream. It first affects the Philippines in October as a weak stream, attaining maximum strength in January. It gradually weakens in March and finally disappears in April. Fig.2 shows the surface air flow over the Philippines in January under the influence of the Northeast Monsoon. Generally this air stream is steady, although at times it may pulsate in surges or outbursts. At other times it is a very weak stream and in extreme cases it may be interrupted by lulls.

The Northeast Monsoon starts as a continental polar air mass with a low temperature of about $-20°C$ and a low humidity with mixing ratio of about 0.5 g/kg near the surface. As it passes over the Pacific, it is transformed into a maritime polar air mass. It finally reaches the Philippines as a maritime tropical air mass, with a surface temperature of about 25°C and a mixing ratio of about 12 g/kg.

This air stream has a moderate temperature inversion at about 1.5 km. Its lapse rate approaches the dry adiabatic below the condensation level. Although its temperature and humidity are only slightly lower than those of other air streams, most of the water content is confined to the lower layer below the inversion while above the inversion it is dry. Table I shows some of the mean thermodynamic properties of the Northeast Monsoon in the vicinity of the Philippines.

The Northeast Monsoon is rather shallow, rarely exceeding $2\frac{1}{2}$ km in depth. Aloft, it is usually overlaid by the Temperate Zone Westerlies over the northern part of Luzon and by the North Pacific Trades over the rest of the Philippines.

The Northeast Monsoon is usually characterized by heavy stratocumulus clouds in the lower levels and showers of heavy drizzle type. Above this deck of low clouds, it is usually devoid of middle and high clouds if dry Temperate Zone Westerlies overlie the Northeast Monsoon. However, if North Pacific Trades lie above the Northeast Monsoon, middle and high clouds are usually present.

TABLE I

MEAN THERMODYNAMIC PROPERTIES OF THE NORTHEAST MONSOON IN THE VICINITY OF THE PHILIPPINES

Height (m)	Temp. (°C)	Specific humidity, q (g water vapour/kg air)	Relative humidity, R.H. (%)	Adiabatic equiv. potential temp., θ_e
surface	25.1	12.2	63	332
500	21.8	11.5	66	332
1,000	18.3	10.2	68	331
1,500	15.5	8.9	68	330
2,000	12.4	8.1	70	329
2,500	9.1	6.5	66	327

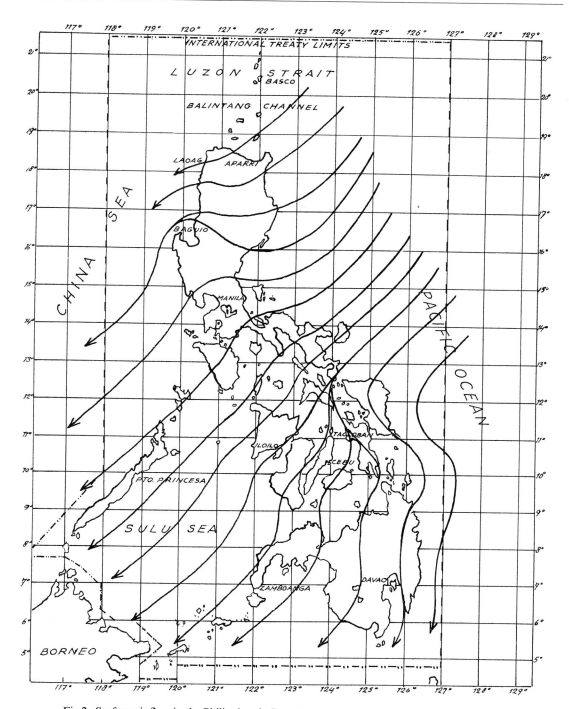

Fig.2. Surface air flow in the Philippines in January.

The Southwest Monsoon

This air stream is also given other names by some authors. Some of these names are Summer Monsoon (RAMAGE, 1959) and Indian Southwesterlies (ESTOQUE, 1952). It originates as Indian Ocean Trades from the Indian Ocean Anticyclone during the Southern Hemisphere winter. Upon crossing the equator, the winds are deflected to the right in

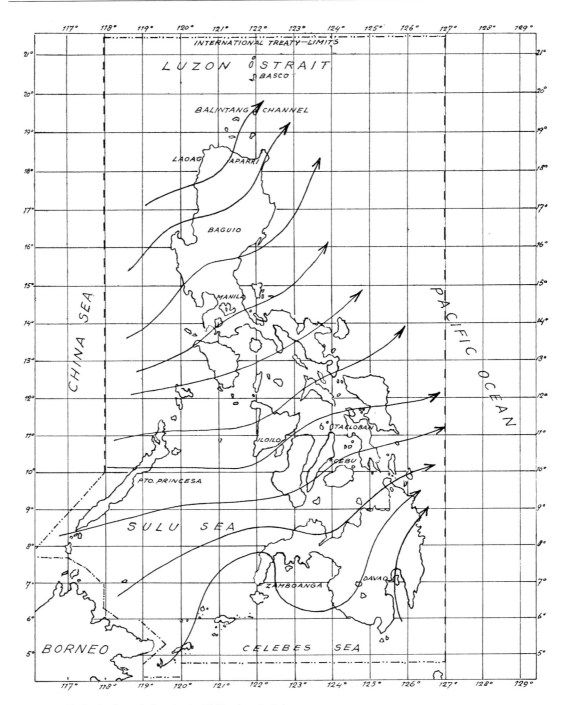

Fig.3. Surface air flow in the Philippines in July.

the Northern Hemisphere, generally arriving in the Philippines as a southwesterly stream, although it may sometimes come from other directions such as westerly or southerly. It may be classified as a maritime equatorial air mass. It is usually of considerable depth, often extending from the earth's surface up to 10 km.

Usually, it first appears in the Philippines in early May, attains maximum intensity in

TABLE II

MEAN THERMODYNAMIC PROPERTIES OF THE SOUTHWEST MONSOON IN THE VICINITY OF THE PHILIPPINES

Height (m)	Mar.–May				June–Aug.				Sept.–Nov.			
	Temp.	q.	R.H.	θ_e	Temp.	q	R.H.	θ_e	Temp.	q	R.H.	θ_e
surface	27.2	19.0	84	356	25.5	17.3	82	350	25.5	17.3	82	350
500	25.1	17.5	83	355	23.4	16.5	85	349	23.7	15.9	81	345
1,000	21.6	14.6	81	347	20.7	15.7	90	348	20.9	13.9	80	343
1,500	19.3	13.3	80	346	18.5	13.4	86	346	18.0	12.6	80	343
2,000	15.9	10.7	77	340	15.6	11.8	83	343	15.1	11.2	80	342
2,500	13.0	9.9	76	338	12.3	9.9	81	340	13.5	10.3	79	342
3,000	10.4	7.6	68	337	9.8	9.1	83	340	10.9	8.4	71	340
4,000	4.5	5.6	64	337	4.0	6.6	82	340	4.6	6.2	71	339
5,000	−2.7	3.8	63	337	−2.0	5.3	85	340	−2.4	4.3	71	337

August and gradually disappears in October. However, occasionally the Southwest Monsoon may appear as early as April over the Philippines and persist up to November or December. Fig.3 shows the dominating influence of the Southwest Monsoon over the Philippines in July.

The Southwest Monsoon is warm and very humid. Its temperature near the surface is generally between 25.5°C and 27.5°C. Its relative humidity is rarely below 70% near the surface. Even at high levels its moisture content is high, and its temperature lapse rate is nearly moist adiabatic with no pronounced inversion. Thus it is quite unstable and is characterized by frequent convective activity. Table II shows some of the mean thermo-dynamic properties of the Southwest Monsoon in the vicinity of the Philippines.

This air stream is also usually constant, although at times it may blow in surges due to the effect of a tropical cyclone or may subside or even be replaced by the North Pacific Trades.

The North Pacific Trades

This air stream is the southern portion of the North Pacific Anticyclone and is therefore classified as a maritime tropical air mass. It travels over a vast expanse of the North Pacific Ocean, arriving at the Philippines from varying directions, generally northeast, east or southeast but sometimes south or even southwest. It is generally dominant over the entire Philippines in April and early May and over the central and southern Philippines in October. It usually overlies the Northeast Monsoon over the eastern sections of the Philippines.

The North Pacific Trades are the warmest air stream that affects the Philippines. They have a temperature of about 27°C near the earth's surface. Their lapse rate is slightly greater than moist adiabatic and they have a weak "trade wind" inversion at about 1.5 km. The moisture content below the inversion is moderate but very dry above the inversion with relative humidity generally below 25%. The air mass is usually conditionally as well as convectively unstable. However, the relatively dry layers at upper levels and the general subsidence within the stream, especially during the period from March to May, prevent the occurrence of intense convective activity. Clouds of the limited convective

TABLE III

MEAN THERMODYNAMIC PROPERTIES OF THE NORTH PACIFIC TRADES IN THE VICINITY OF THE PHILIPPINES

Height (m)	Dec.–Feb.				Mar.–May				June–Aug.				Sept.–Nov.			
	Temp.	q	R.H.	θ_e	Temp.	q	R.H.	θ_e	Temp.	q	R.H.	θ_e	Temp.	q	R.H.	θ_e
surface	26.6	13.5	59	339	28.0	14.5	60	343	27.2	16.0	68	347	27.0	17.1	74	350
500	23.8	12.2	61	336	25.5	13.8	66	342	24.6	14.7	70	345	24.1	15.4	77	345
1,000	19.3	10.9	69	333	20.8	12.4	73	339	20.7	13.5	79	343	20.5	13.4	79	342
1,500	15.7	10.2	76	332	17.5	11.6	77	339	17.7	12.4	83	340	18.2	12.2	79	340
2,000	13.1	8.0	67	330	14.5	9.3	71	335	14.5	9.8	77	335	15.7	10.8	79	339
2,500	11.1	7.0	64	330	12.0	7.1	60	332	12.3	7.9	66	334	13.1	9.0	71	339
3,000	8.6	5.8	56	330	9.9	5.6	52	330	9.5	6.5	62	332	10.6	7.4	64	337
4,000	4.0	3.4	42	330	4.1	3.5	42	330	3.5	4.4	55	332	4.4	5.8	64	336
5,000	−2.8	2.2	38	330	−2.1	2.3	38	331	−2.0	3.4	56	335	−2.5	3.9	66	336

types, fair weather cumulus and stratocumulus, prevail in this air stream, except, of course, in areas affected by topographic convergence where towering cumulus and showers may develop. On the other hand, especially during the months from June to December when the North Pacific Anticyclone is weak, general convergence frequently takes place within this air stream and the inversion is destroyed. This results in increased convection with moisture being transported up to great heights. This, in turn, results in the formation of cumulus congestus or even cumulonimbus clouds in the afternoons. Some of the mean thermodynamic properties of the North Pacific Trades in the vicinity of the Philippines are shown in Table III.

The Temperate Zone Westerlies

This air stream exists mainly at upper levels over the Philippines, overlying the Northeast Monsoon or the Southwest Monsoon especially over northern Luzon. The Temperate Zone Westerlies start appearing over the Batanes in October above 5 km, gradually increasing in strength and extending to about 13°N latitude up to April. They are then gradually pushed northward by the Southwest Monsoon. Practically no trace of this air stream is found over the Philippines from July to September. This air stream is warm and moist and its thermodynamic properties are similar to those of the Southwest Monsoon.

The South Pacific Trades

This air stream comes from the South Pacific Anticyclone. It follows a southeasterly or easterly trajectory across the East Indies, gradually veers as it crosses the equator and reaches the Philippines from the south or southwest. It appears over the southern Philippines during the period from May to July, but it is often difficult to distinguish it from the Southwest Monsoon.
The South Pacific Trades are warm and quite moist although they are relatively dry at upper levels. They are also frequently characterized by a weak "trade wind" inversion at about 1.5 km. This air stream may be classified thermodynamically between the maritime tropical air mass and the maritime equatorial air mass.

Ocean currents

A knowledge of the characteristics of the ocean currents around the Philippines is necessary for understanding its climate. Oceanographers (SVERDRUP, 1943) found that the
main current affecting the Philippines is the North Equatorial Current moving westward
across the North Pacific Ocean. It splits into northward and southward branches upon
reaching the eastern coastal areas of the Philippines. The northward branch flows along
the east coasts of northeastern Visayas and Luzon and becomes the Kuroshio Current.
The southern branch flows along the east coasts of southern Visayas and Mindanao
and recurves to the east as the Equatorial Counter Current. On the western coasts of
Visayas and Mindanao there is a southward or southeastward current, while on the
western coast of Luzon the prevailing current is northward.

The surface temperature of the seas around the Philippines is relatively high. A study of
more than 3,800 observations of ships from December 1920 to April 1931 (SELGA et al.,
1931) showed that the average temperature of the sea surface in the vicinity of the Philippines between 4°N and 23°N and 117°E and 127°E was 27.3°C, as compared to the
average air temperature of 26.9°C for the same period for landstations near sea level. It
was also found that the temperature of the sea surface in Philippine waters is quite uniform
and that the annual range of temperature variations is very small. An annual range of
7°C occurs in the Bashi and Balintang Channels north of Luzon, 5°C in the South China
Sea and the Pacific Ocean bordering the western and eastern coasts of Luzon, and a
variation not exceeding 4°C for all of the seas south of Luzon. These values are in general
agreement with the values indicated in the *Climatological and Oceanographic Atlas for
Mariners* (U.S. DEPARTMENT OF COMMERCE AND HYDROGRAPHIC OFFICE, 1961).

Linear systems

The Philippines is affected to a certain degree by linear systems, the most important of
which are fronts, the Intertropical Convergence Zone, and easterly waves. The characteristics and behaviour of these systems are described below.

Fronts

The Philippines is affected by fronts during the winter months only. Global charts (TRE-
WARTHA, 1954) show the polar front in January oriented west-southwest to east-northeast
across the northern tip of Luzon. Other charts (HAURWITZ and AUSTIN, 1944) show the
polar front in January across southern Luzon oriented southwest to northeast. Other
authors (DEPPERMAN, 1936) refer to this front as the tropical front and locate it over the
Balintang Channel between Luzon and Formosa in a southwesterly to northeasterly
orientation. These are, of course, mean positions, since the front is constantly on the
move from north to south or south to north, at times even reaching Mindanao. This
front is considered as the interface between the Northeast Monsoon and the North
Pacific Trades.

However, most investigators found this front to be very weak since the two adjoining
air masses have almost identical characteristics in the vicinity of the Philippines. ESTOQUE
(1952), after investigating all available soundings for 1950 and 1951, found no true frontal

inversion between the Northeast Monsoon and the North Pacific Trades. Although there is unmistakable kinematic evidence that a portion of the North Pacific Trades often overlies the Northeast Monsoon, the surface separation between the two air streams does not have the characteristics of a true front. The moisture and temperature inversions which indicate overrunning of the warmer over the colder and drier air are absent. Nevertheless, this front, coupled with topographic effects, is responsible for a portion of the rainfall and cloudiness along the eastern coasts of the Philippines and occasionally even over the middle and western sections of the islands.

The Intertropical Convergence Zone

This zone is given other names such as Intertropical Front, Equatorial Trough, Near Equatorial Trough, Doldrum Belt, etc. These terms are used to designate the zone of discontinuity in the wind field between a stream from the Northern Hemisphere and another stream from the Southern Hemisphere.

The Intertropical Convergence Zone affects the Philippines from May to October. From November to April it is south of the Philippines. It starts appearing in the southwestern portion of the Philippines in May, moves northward and reaches its northernmost position in July or August when it is well to the north of the Philippines. In early August it moves southward and is well to the south of the Philippines in November. It reaches its southernmost position in January or February, then it moves northward again. Fig.4 shows the mean monthly positions of the Intertropical Convergence Zone in the vicinity of the Philippines. The variations in the position of the Intertropical Convergence Zone from the mean in the vicinity of the Philippines are quite large and disorganized. Occasionally there are rapid surges northward or southward, depending on the relative strengths of the air streams on either side of the zone and also on the movement of tropical cyclones along it. The Intertropical Convergence Zone is usually characterized by disturbed weather conditions consisting of widespread cloudiness, precipitation, and moderate to strong surface winds associated mostly with mesoscale systems. However, weather conditions along the Intertropical Convergence Zone vary from day to day. Some investigators (ESTOQUE, 1952) found poor correlation between convergence and disturbed weather in this zone. Precipitation in this zone is in the general nature of the convective type and is seldom continuous, except when the zone is very well defined. Since the temperature difference between the two air masses along the zone is very small, most investigators are loath to call this a frontal zone.

Easterly waves

Easterly waves are referred to as wavelike perturbations which move from east to west in the easterly currents of the tropics. Waves of this type affect the Philippines about twice per week in varying intensities. They are more frequent in the summer and seldom occur during the winter months.

These easterly waves are usually accompanied by cloudiness and precipitation, especially along the orographic obstacles on the eastern coastal areas of the Philippines. At times their only effect may be an increase in cloudiness, but at other times heavy showers or even continuous rains may persist for 2–3 days.

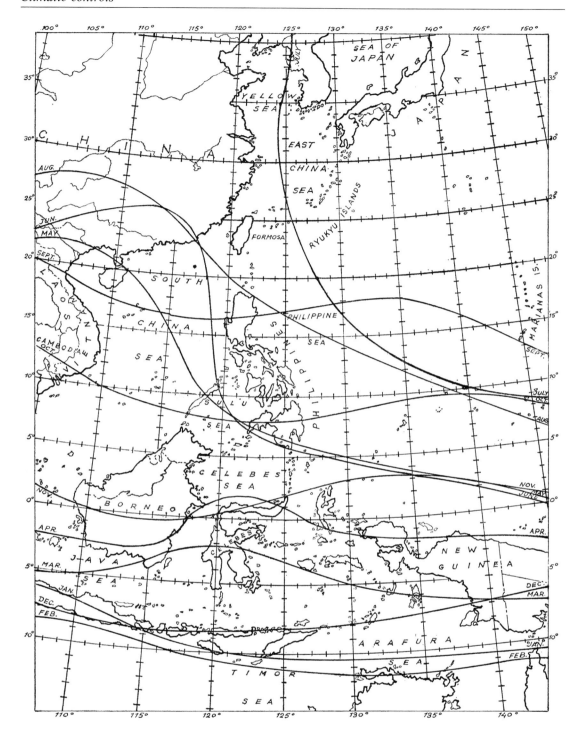

Fig.4. Mean monthly positions of the Intertropical Convergence Zone in the vicinity of the Philippines.

Tropical cyclones

No assessment of the climate of the Philippines can be complete without a discussion of the effects of tropical cyclones. Although these cyclones cause enormous losses to life and property in general, they also have some beneficial effects. These cyclones contribute largely to the rainfall in the Philippines from May to December. They affect prevailing winds, humidity and cloudiness, and are usually responsible for the maximal values of rainfall, strongest winds and minimal pressures observed in many places. They also influence the behaviour of other climatic elements to a certain degree.

The Philippines are located in a region which is recognized as having the greatest frequency of tropical cyclones in the world. Frequency of tropical cyclones in 70 years (1884–1953) for the western Pacific Ocean and China Sea area bounded by latitudes 5°N–30°N and longitudes 105°E–150°E was about 1,541, or an annual mean of about 22 (Chin, 1958). An investigation of all tropical cyclones which affected the Philippines for the 15 year period from 1948 to 1962 showed an annual mean of 19.3. This is midway between the results of a similar investigation made by Coronas (1920) for the 11 year period 1908–1918 wherein he obtained an annual mean of 17.3, and an investigation by Algue (1904) for the 22 year period 1880–1901 wherein he obtained an annual mean of 21.3. For the period 1948–1962, the annual number of tropical cyclones which affected the Philippines varied from a minimum of 13 in 1950 and 1951 to a maximum of 29 in 1952.

The tropical cyclone season in the Philippines lasts from June to December, although the other months are not entirely free of these cyclones. Based on the records for 1948–

TABLE IV

MEAN MONTHLY AND ANNUAL FREQUENCY OF TROPICAL CYCLONES AFFECTING THE PHILIPPINES, 1948–1962

Year	Jan.	Feb.	Mar.	Apr.	May	June	July	Aug.	Sep.	Oct.	Nov.	Dec.	Annual summary
1948	1	0	0	0	2	0	3	1	3	2	6	3	21
1949	1	0	0	0	0	3	4	2	4	3	4	1	22
1950	0	0	0	0	0	2	3	1	3	2	1	1	13
1951	0	0	0	0	0	1	1	4	2	1	2	1	13
1952	0	0	0	0	1	5	2	4	4	5	3	5	29
1953	0	1	0	0	1	2	0	5	2	2	4	1	18
1954	0	0	1	0	1	0	1	6	2	3	3	1	18
1955	1	1	0	1	0	0	2	3	1	4	1	1	15
1956	0	0	1	2	0	0	5	4	5	1	5	3	26
1957	2	0	0	1	0	2	1	2	3	3	1	0	15
1958	1	0	0	0	0	1	4	3	3	2	4	0	18
1959	0	1	1	0	0	0	1	4	2	4	3	2	18
1960	1	0	0	1	1	2	2	6	1	3	0	2	19
1961	1	1	1	0	1	3	4	4	4	1	1	2	23
1962	0	1	0	0	2	0	5	6	4	1	3	0	22
Total	8	5	4	6	9	21	38	55	43	37	41	23	290
Rank	9	11	12	10	8	7	4	1	2	5	3	6	
Median	0	0	0	0	0	1	2	4	3	2	3	1	18
Mode	0	0	0	0	0	0	1	4	3	2	2	1	18
Average	0.5	0.3	0.3	0.4	0.6	1.4	2.5	3.7	2.9	2.5	2.7	1.5	19.3
Percentage of annual Total	2.8	1.7	1.4	2.1	3.1	7.2	13.1	19.0	14.8	12.8	14.1	7.9	

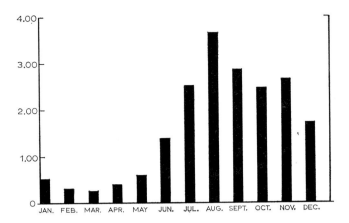

Fig.5. Mean monthly frequency of tropical cyclones affecting the Philippines.

1962, the period from June to December accounts for about 89% of the mean annual total number of tropical cyclones affecting the Philippines. August has the greatest mean frequency with 3.7 and September is second with 2.9. March and February have the smallest mean frequencies with about 0.3. Table IV shows the mean monthly and annual frequency distribution of tropical cyclones that affected the Philippines from 1948 to 1962. These statistics are not very different from the results obtained by ALGUE (1904) and CORONAS (1920). Fig.5 is a graph showing the monthly frequency distribution of tropical cyclones affecting the Philippines.

Tropical cyclones follow widely variable tracks in the vicinity of the Philippines. During the months of April, May and June, the cyclones which hit the Philippines generally cross the Visayas. During the months of July, August and September, most of the cyclones cross northern Luzon or the Batanes Islands. Again, during the period from October to March, cyclones generally cross the Visayas. Mean monthly tracks of tropical cyclones in the vicinity of the Philippines are shown in Fig.6.

Percentage frequency of tropical cyclone passage over northeastern Luzon, the Batanes, Northern Samar, Sorsogon, and Masbate is between 31–40%; 21–30% pass through northwestern Luzon, most of southern Luzon, and northern Leyte; 11–20% pass through central Luzon and a portion of northern and southern Luzon, and the greater portion of the Visayas. Less than 10% pass through Mindanao. Fig.7 shows the percentage frequency distribution of cyclone passage through the different parts of the Philippines.

Tropical cyclones in the vicinity of the Philippines move at widely varying speeds ranging from less than 2 m/sec to more than 10 m/sec, but the average speed is about 6 m/sec. Individual cyclones affect the Philippines from 1 to 7 days, depending on their tracks and speeds of movement.

Most of the statistics on tropical cyclones used in this discussion are based on records for the period 1948–1962. These records were selected due to the high confidence which can be placed on analyses of positions of the cyclone centers due to numerous reports from reconnaissance aircraft, ship and land stations.

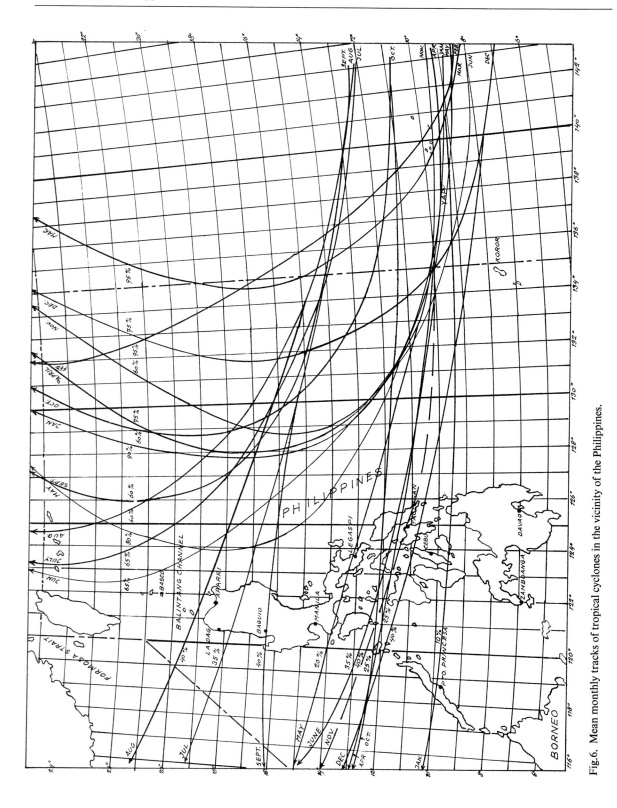

Fig.6. Mean monthly tracks of tropical cyclones in the vicinity of the Philippines.

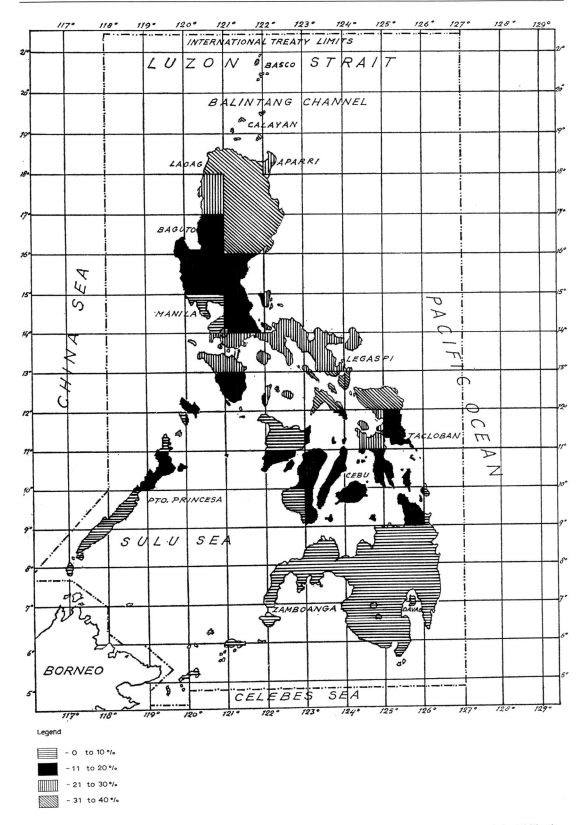

Legend

- 0 to 10 %
- 11 to 20 %
- 21 to 30 %
- 31 to 40 %

Fig.7. Mean percentage frequencies of tropical cyclone passage in the different parts of the Philippines.

Thunderstorms

Compared to tropical cyclones, thunderstorms are relatively small and short period disturbances. Nevertheless, owing to the moist and unstable conditions of most of the air streams affecting the Philippines combined with the aid of orographic lifting, thunderstorms occur frequently throughout the Philippines during practically all months of the

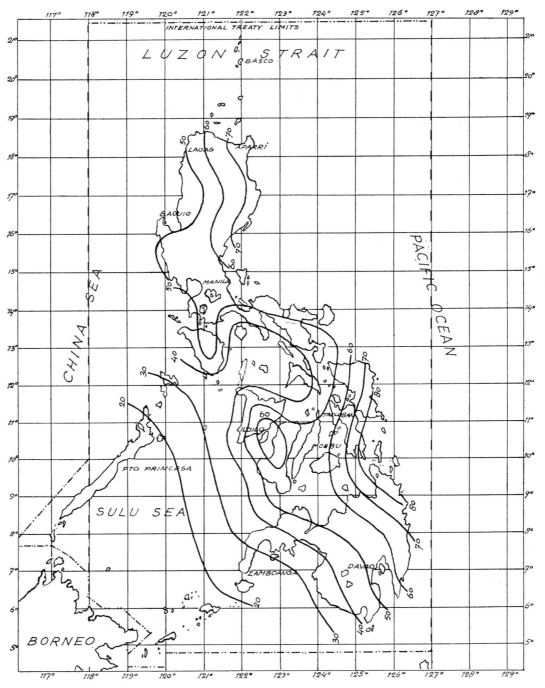

Fig.8. Distribution of mean annual number of days with thunderstorms in the Philippines.

year. Fig.8 shows the distribution of the mean annual number of days with thunderstorms in the Philippines, while Fig.9 is a graph showing the monthly distribution of thunderstorm days for ten selected stations in the Philippines. Generally, effects of thunderstorms on the climatic elements are not very pronounced, but in some instances the rainfall accompanying them may be considerable.

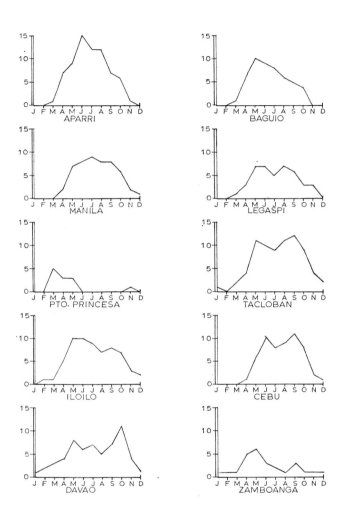

Fig.9. Mean monthly distribution of thunderstorm days for ten selected stations in the Philippines.

Climatic elements

The characteristics and behaviour of the climatic controls affecting the Philippines have notable effects on the climatic elements. In the Philippines, as elsewhere, the most important of the climatic elements are temperature, rainfall, humidity, cloudiness, and winds. Atmospheric pressure, on the other hand, is considered to be of minor importance. The character and patterns of each of these elements are presented below.

Temperature

The Philippines, situated in the tropics and consequently in a region of high insolation, surrounded by warm seas, and with warm air currents flowing over them, is expected to have generally high temperatures. Table V shows the mean monthly and annual temperatures for 44 stations in the Philippines. From this table, the following important points can be seen:

Mean annual temperatures

Taking into account only the 42 stations with elevations of less than 50 m (Baguio with an elevation of 1,482 m and Malaybalay with an elevation of 642 m are excluded), the mean annual temperature of the Philippines is 27.0°C. The average of annual temperatures for 19 stations in Luzon is 26.8°C as compared to 27.3°C for 14 stations in the Visayas and 26.9°C for 9 stations in Mindanao. Mean annual temperatures for the stations with elevations of less than 50 m range from 25.7°C for Basco, the northernmost station, to 27.9°C for Masbate in the middle of the Philippines. The effect of elevation can easily be seen by comparing the mean annual temperatures of 18.2°C for Baguio and 23.6°C for Malaybalay with the mean annual values for the other stations.

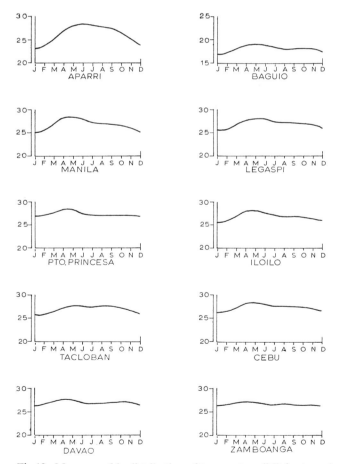

Fig.10. Mean monthly distribution of temperature (°C) for ten selected stations in the Philippines.

TABLE V

MEAN MONTHLY AND ANNUAL TEMPERATURES (°C) FOR 44 STATIONS IN THE PHILIPPINES

Station	Years	Jan.	Feb.	Mar.	Apr.	May	June	July	Aug.	Sep.	Oct.	Nov.	Dec.	Annual
Luzon														
Ambulong	41	25.5	26.1	27.3	28.5	28.7	27.8	27.1	27.0	26.8	26.5	26.4	25.6	26.9
Aparri	52	23.2	23.8	25.3	27.1	28.1	28.5	28.1	27.9	27.5	26.5	25.2	23.9	26.3
Aurora	12	25.4	25.7	26.3	27.3	27.9	27.8	27.3	27.3	27.1	26.8	26.5	25.7	26.8
Baguio	52	16.9	17.4	18.3	18.9	19.2	18.9	18.4	18.0	18.2	18.3	18.1	17.5	18.2
Baler	26	24.5	24.7	25.6	26.9	27.9	28.2	28.0	28.1	27.7	27.0	26.1	25.2	26.7
Basco	51	22.3	22.7	23.9	25.1	27.6	28.3	28.3	27.9	27.5	26.3	24.8	23.1	25.7
Cabanatuan	14	25.9	26.6	27.8	29.3	29.8	28.7	28.1	27.6	27.7	27.6	26.7	26.0	27.7
Casiguran	12	23.7	24.0	24.9	26.4	27.2	27.6	27.4	27.2	27.0	26.5	25.5	24.3	26.0
Daet	34	25.1	25.2	25.9	26.9	27.7	27.9	27.4	27.4	27.1	26.6	26.3	25.5	26.6
Dagupan	52	25.8	26.4	27.7	29.0	29.0	28.2	27.4	27.3	27.4	27.5	26.8	26.1	27.4
Iba	23	25.4	25.7	26.8	28.1	28.4	27.7	27.0	26.7	26.8	27.0	26.6	25.9	26.8
Infanta	24	24.6	25.0	25.9	27.1	27.9	28.4	28.0	28.0	27.6	26.9	26.3	25.3	26.8
Laoag	46	24.7	25.3	26.9	28.3	28.6	27.9	27.3	27.0	27.2	26.9	26.2	25.2	26.8
Legaspi	52	25.7	25.9	26.7	27.7	28.2	28.1	27.4	27.4	27.3	27.1	26.7	26.2	27.0
Lucena	24	25.3	25.8	26.7	28.0	28.7	28.3	27.8	27.7	27.6	27.0	26.5	25.5	27.1
Manila CO	71	25.0	25.5	26.8	28.3	28.6	27.9	27.1	27.0	26.9	26.7	25.9	25.2	26.7
Tuguegarao	51	23.4	24.4	26.4	28.2	29.0	28.9	28.2	27.9	27.5	26.4	25.1	24.0	26.6
Vigan	52	25.4	25.7	27.0	28.3	28.7	28.0	27.3	26.9	27.1	27.3	26.7	26.0	27.0
Virac	25	25.9	25.9	26.4	27.2	27.8	28.1	27.8	28.1	27.8	27.3	26.9	26.4	27.1
Manilla MMO	14	25.4	26.1	27.2	28.9	29.4	28.5	27.7	27.3	27.4	27.1	26.2	25.5	27.2
Average for Luzon[1]		24.9	25.3	26.4	27.7	28.4	28.1	27.6	27.5	27.3	26.9	26.2	25.3	26.8
Visayas														
Borongan	25	25.9	25.8	26.4	27.1	27.6	27.6	27.6	27.8	27.8	27.1	26.8	26.3	27.0
Calapan	22	25.5	25.8	26.8	27.9	28.1	27.7	27.3	27.2	27.2	27.0	26.6	25.9	26.9
Catarman	12	25.4	25.4	25.9	26.6	27.4	27.6	27.3	27.6	27.4	26.8	26.5	25.8	26.6
Catbalogan	24	25.9	26.1	26.7	27.6	28.1	28.1	27.9	28.2	27.9	27.5	26.8	26.3	27.3
Cebu	52	26.3	26.4	27.2	28.2	28.4	28.0	27.6	27.7	27.6	27.5	27.1	26.7	27.4
Coron	11	27.1	27.3	28.0	28.9	28.6	27.7	26.7	26.9	27.0	27.4	27.4	27.0	27.5
Cuyo	27	26.9	27.0	27.9	28.8	28.7	28.0	27.4	27.3	27.3	27.6	27.6	27.3	27.7
Dumaguete	23	26.4	26.5	27.2	27.9	28.0	27.7	27.4	27.7	27.5	27.3	27.1	26.8	27.3
Iloilo	52	25.8	26.1	26.9	28.2	28.3	27.7	27.2	26.9	27.1	26.9	26.6	26.1	27.0
Masbate	25	26.3	26.8	27.4	28.6	29.3	29.1	28.7	28.4	28.3	28.2	27.5	26.7	27.9
Pto. Princesa	10	26.9	27.2	27.7	28.5	28.5	27.4	27.2	27.1	27.1	27.2	27.2	26.9	27.4
Romblon	28	26.4	26.5	27.6	28.8	29.0	28.7	28.2	28.0	28.1	27.8	27.4	26.7	27.8
Roxas	52	26.0	26.2	26.9	27.9	28.4	27.8	27.4	27.4	27.1	27.2	27.0	26.4	27.1
Tacloban	51	25.7	25.8	26.4	27.3	27.8	27.6	27.5	27.7	27.6	27.2	26.7	26.0	26.9
Average for Visayas		26.2	26.4	27.1	28.0	28.3	27.9	27.5	27.6	27.5	27.3	27.0	26.5	27.3
Mindanao														
Cagayan de Oro	24	25.9	26.2	26.8	27.6	28.0	27.7	27.2	27.4	27.2	27.2	26.8	26.2	27.0
Cotabato	22	27.1	27.4	27.8	28.3	28.0	27.7	27.2	27.3	27.4	27.6	27.5	27.2	27.5
Davao	30	26.3	26.6	27.2	27.7	27.6	27.0	26.8	26.9	27.0	27.1	27.0	26.6	27.0
Dipolog	12	26.7	27.0	27.5	28.1	28.0	27.7	27.3	27.4	27.2	27.4	27.3	26.8	27.4
General Santos	12	26.6	27.0	27.5	27.8	27.4	26.7	26.2	26.2	26.5	26.7	26.9	26.8	26.9
Hinatuan	11	25.6	25.5	25.9	26.4	26.9	26.8	26.8	27.1	27.0	26.9	26.7	25.9	26.5
Jolo	29	26.2	26.0	26.3	26.7	27.0	26.9	26.9	27.0	26.9	26.6	26.5	26.4	26.6
Malaybalay	12	22.9	23.1	23.6	24.2	24.4	24.1	23.5	23.5	23.6	23.7	23.6	23.1	23.6
Surigao	51	25.6	25.6	26.1	26.7	27.2	27.4	27.4	27.7	27.5	27.1	26.4	25.8	26.7
Zamboanga	51	26.3	26.4	26.8	27.1	27.1	26.8	26.6	26.8	26.7	26.6	26.6	26.4	26.7
Average for Mindanao[2]		26.3	26.4	26.9	27.4	27.5	27.2	26.9	27.1	27.0	27.0	26.9	26.4	26.9
Average for Philippines[3]		25.6	25.9	26.7	27.7	28.2	27.9	27.4	27.4	27.3	27.1	26.6	25.9	27.0

[1] Baguio not included
[2] Malaybalay not included
[3] Baguio and Malaybalay not included

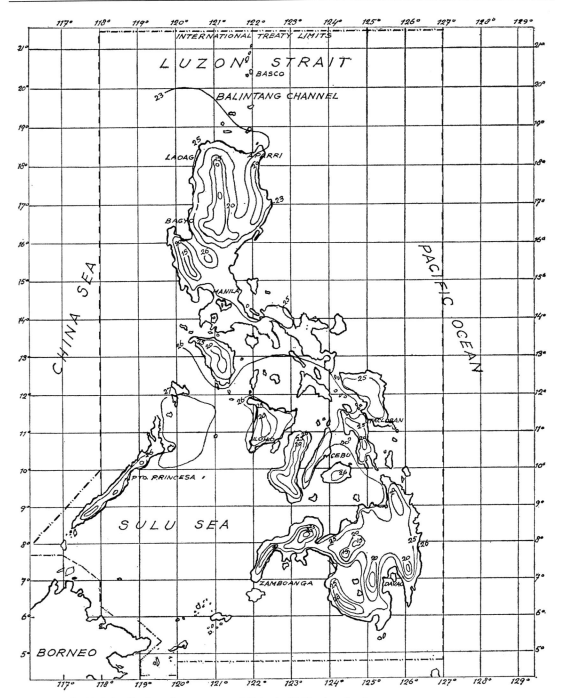

Fig.11. Distribution of temperature (°C) in the Philippines for January.

Monthly distribution and annual range of temperature

For the whole country in general, the hottest months are May with 28.2°C, June with 27.9°C and April with 27.7°C. The coldest months are January with 25.6°C, February with 25.9°C and December with 25.9°C. The average of annual ranges for 42 stations is 2.6°C. By regions, the average annual range for 19 stations in Luzon is 3.6°C, for 14

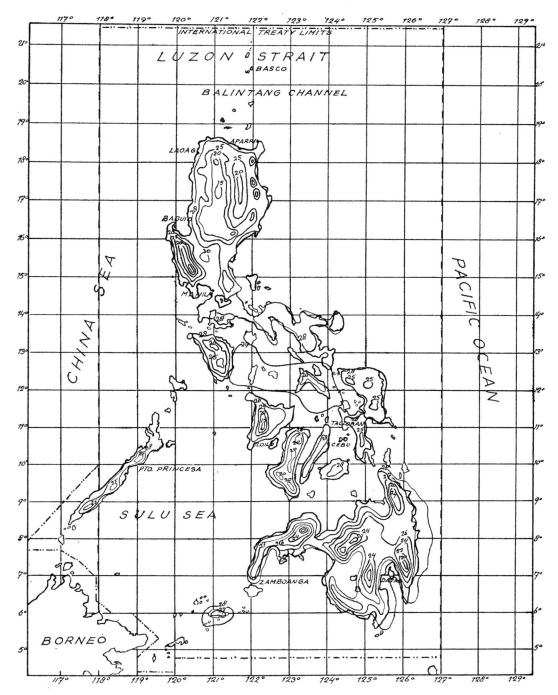

Fig.12. Distribution of temperature (°C) in the Philippines for May.

stations in the Visayas the average annual range is 2.2°C, and for 9 stations in Mindanao the average annual range is 1.4°C. Generally, there is a decrease of the annual range values from the most northerly station to the southernmost stations. Basco, the northernmost station, has an annual range of 6.0°C as compared to 0.8°C for Zamboanga, one of the southernmost stations. Fig.11–12 show the temperature distribution in the Philippines during the months of January and May, the coldest and hottest months, respectively.

Hourly distribution and diurnal range of temperature

The highest temperatures for most places in the Philippines occur between 1 p.m. and 3 p.m. while the lowest temperatures occur between 5 a.m. and 7 a.m. The diurnal range of temperature for most stations in the Philippines is much larger than the annual range. The average of diurnal ranges for sixteen selected stations in the Philippines is 7.5°C which is about 3 times the average annual range of 2.6°C. Cagayan de Oro has the largest diurnal range of 9.6°C while Cebu has the smallest with 6.7°C. In other words, there is a difference of only 2.9°C between the largest and smallest values.

Extremes of temperature

The absolute maximum temperature recorded in the Philippines was 42.2°C at Tuguegarao, situated in a valley in northern Luzon, on 29 April 1912. The absolute minimum of 3.0°C was recorded at Baguio, at an elevation of 1,482 m, during January 1903. In general, the highest temperatures are observed in the valleys and plains of Luzon, while the lowest temperatures occur at stations with high elevations.

Rainfall

This element is one of the most important and most interesting in the discussion of the climate of the Philippines. The rainfall of the Philippines is influenced to a large extent by the air streams, tropical cyclones, the Intertropical Convergence Zone and topography. It is influenced to a lesser extent by fronts, easterly waves and local thunderstorms. Table VI shows the mean monthly and annual rainfall distribution for 44 stations in the Philippines. From this table, the following important points can be gathered:

Mean annual rainfall

The average annual rainfall for the Philippines is 2,533.4 mm. Luzon has an annual average of 2,724.4 mm, Visayas has 2,391.7 mm, and Mindanao has 2,349.8 mm. Hinatuan, on the eastern coast of Mindanao, has the largest annual mean of 4,305.2 mm, while General Santos, situated in a valley in southern Mindanao, has the lowest annual mean of 933.8 mm. Fig.13 shows the distribution of annual rainfall in the Philippines. Of the 44 stations whose rainfall records were studied, only one had an annual mean of less than 1,000 mm. Thirteen had annual means between 1,000 and 2,000 mm, 17 between 2,000 and 3,000 mm, 10 between 3,000 and 4,000 mm, and 3 had more than 4,000 mm. Of the 13 stations having annual means greater than 3,000 mm, 11 were directly exposed to the Northeast Monsoon while only 2 were exposed to the Southwest Monsoon. This may be in contradiction with results obtained by other investigators who arrived at the conclusion that the western coastal areas receive more rainfall than the eastern coastal areas owing to the fact that the Southwest Monsoon is more humid than the Northeast Monsoon. Upon closer scrutiny, it is evident that most of the stations along the eastern coasts of the islands do not derive their rainfall only from the Northeast Monsoon, but also from the Southwest Monsoon and the North Pacific Trades, especially with the additional help of tropical cyclones. This observation is supported by the fact that many of the eastern coastal

TABLE VI

MEAN MONTHLY AND ANNUAL RAINFALL (MM) FOR 44 STATIONS IN THE PHILIPPINES

Station	Years	Jan.	Feb.	Mar.	Apr.	May	June	July	Aug.	Sep.	Oct.	Nov.	Dec.	Annual
Luzon														
Ambulong	41	6.4	14.0	20.6	47.2	147.1	199.6	305.8	338.8	299.0	232.2	168.4	111.0	1910.1
Aparri	52	144.0	89.7	54.6	48.5	111.0	172.5	189.5	234.2	295.1	366.8	335.8	217.9	2259.6
Aurora	12	58.2	21.1	44.5	30.5	94.5	153.7	241.8	212.6	212.3	261.6	201.9	166.9	1699.6
Baguio	52	19.6	19.1	48.3	121.7	351.8	422.1	921.5	1042.9	649.7	361.7	162.6	55.9	4176.9
Baler	50	201.9	176.8	221.0	285.0	286.8	231.1	258.1	198.9	288.5	375.2	403.6	378.5	3305.4
Basco	51	229.4	129.5	128.0	97.8	205.5	241.8	307.6	402.3	363.5	366.8	345.7	324.1	3142.0
Cabanatuan	15	5.6	6.4	25.7	52.6	125.2	210.8	276.4	410.5	304.0	175.5	129.5	76.7	1798.9
Casiguran	14	230.6	224.5	268.5	166.1	201.4	257.8	215.6	244.6	323.3	377.2	579.4	528.8	3617.8
Daet	34	382.0	249.9	212.3	152.4	162.6	166.1	224.5	198.1	298.7	540.8	660.1	546.3	3793.8
Dagupan	52	8.9	13.0	25.4	84.6	202.9	315.0	534.7	541.2	436.1	170.7	80.5	25.7	2438.7
Iba	45	6.4	5.1	17.8	44.2	256.5	568.7	904.7	984.8	686.8	232.2	85.3	33.0	3825.5
Infanta	28	404.1	262.9	190.0	200.2	212.6	208.0	229.1	232.4	280.9	502.2	555.8	519.2	3797.4
Laoag	46	4.3	5.1	4.6	11.4	178.3	313.2	531.9	627.9	388.1	168.1	47.8	13.5	2294.2
Legaspi	52	366.0	264.9	218.2	158.0	178.3	194.1	235.0	209.3	251.7	313.4	478.8	502.9	3370.6
Lucena	33	113.0	55.4	50.0	56.9	107.2	168.9	160.8	182.1	184.9	266.7	310.9	256.5	1913.3
Manila CO	90	22.9	10.9	16.8	32.3	128.3	252.5	413.5	436.9	353.1	195.3	137.9	68.3	2068.7
Tuguegarao	51	31.2	24.9	34.0	64.5	131.8	156.2	232.9	206.2	236.2	237.5	269.0	138.7	1763.1
Vigan	52	3.0	6.1	8.4	20.3	186.7	369.1	651.0	710.2	400.8	140.0	43.7	13.2	2552.5
Virac	46	241.0	191.0	160.5	126.0	174.0	201.2	228.6	166.6	246.1	366.0	479.6	425.7	3006.3
Manila MMO	14	15.5	6.6	8.4	19.8	99.1	216.2	263.7	464.8	274.1	190.2	120.4	74.9	1753.7
Average for Luzon		125.7	88.8	87.9	91.0	177.1	250.9	366.3	402.3	338.7	292.0	279.8	223.9	2724.4
Visayas														
Borongan	52	625.9	436.9	359.4	270.3	274.8	228.9	191.5	156.7	181.1	340.1	543.1	653.8	4262.5
Calapan	36	111.3	69.9	58.9	109.7	176.5	215.1	238.8	222.8	193.0	277.9	289.3	198.6	2161.8
Catarman	12	418.8	233.2	259.3	160.5	136.4	143.8	175.5	182.9	189.0	447.0	584.5	515.9	3446.8
Catbalogan	39	281.2	184.2	159.0	135.9	165.1	201.2	247.9	213.9	256.3	318.5	363.7	375.4	2902.3
Cebu	52	104.9	71.1	54.6	52.8	120.9	177.0	196.6	152.7	186.9	201.4	162.8	137.7	1619.4
Coron	11	37.6	6.6	7.6	30.0	168.9	366.8	376.2	496.1	451.9	289.6	140.0	126.5	2497.8
Cuyo	52	17.3	8.4	7.9	32.5	197.6	336.3	424.2	386.8	368.0	284.5	152.1	74.7	2290.3
Dumaguete	43	97.8	70.1	55.4	47.0	110.5	149.1	148.3	113.0	135.9	188.7	175.5	137.9	1429.2
Iloilo	52	59.2	38.1	35.6	51.6	153.4	264.7	389.6	370.3	293.6	262.9	206.5	120.9	2246.4
Masbate	50	185.9	114.0	80.8	39.4	102.4	153.9	180.3	170.7	176.0	192.3	224.3	258.8	1878.8
Pto. Princesa	14	37.8	28.2	60.2	42.2	135.6	170.7	181.1	201.4	220.7	206.8	204.2	118.4	1607.3
Romblon	51	117.3	78.2	64.0	72.1	142.2	209.8	276.9	227.6	230.4	321.1	297.4	234.2	2271.2
Roxas	52	134.9	82.3	54.1	57.9	176.8	262.4	268.7	229.9	249.4	368.8	290.1	226.1	2401.4
Tacloban	51	318.8	209.0	177.8	133.9	163.3	172.0	160.5	133.1	150.9	207.3	288.3	353.3	2468.2
Average for Visayas		182.0	116.4	102.5	88.3	158.9	218.0	246.9	232.7	234.5	279.1	280.1	252.3	2391.7
Mindanao														
Cagayan de Oro	45	90.2	54.4	45.2	34.8	127.5	212.1	234.2	192.8	216.7	193.3	135.1	124.5	1660.8
Cotabato	43	80.8	85.1	84.1	150.1	222.0	239.5	266.7	242.8	225.0	280.9	183.6	103.9	2164.5
Davao	52	118.1	103.4	119.1	142.0	235.7	216.7	175.5	161.5	175.5	192.5	143.3	145.0	1928.3
Dipolog	12	75.2	53.1	84.1	101.3	217.2	312.2	384.3	431.0	295.9	431.0	313.7	222.3	2921.3
General Santos	12	43.7	56.6	49.0	52.1	112.8	110.7	99.6	105.7	76.2	78.5	81.8	67.1	933.8
Hinatuan	22	723.9	426.7	460.0	322.3	294.9	220.0	217.9	198.6	195.3	209.6	371.3	664.7	4305.2
Jolo	57	104.6	106.2	107.2	143.0	217.2	223.0	178.3	170.2	188.0	226.3	215.1	169.2	2048.3
Malaybalay	12	107.2	67.8	92.5	95.5	244.1	339.1	357.9	339.9	370.1	331.5	195.3	195.3	2736.2
Surigao	52	554.0	381.5	367.8	256.0	176.3	122.2	173.5	136.7	169.4	278.9	425.7	634.0	3676.0
Zamboanga	54	51.8	52.6	38.4	52.6	89.7	112.8	126.0	113.0	122.9	155.4	120.7	88.1	1124.0
Average for Mindanao		195.0	138.7	144.7	135.0	193.7	210.8	221.4	209.2	203.5	237.8	218.6	241.4	2349.8
Average for Philippines		159.4	109.0	105.4	100.1	175.1	231.3	295.4	304.4	274.8	275.6	266.0	236.9	2533.4

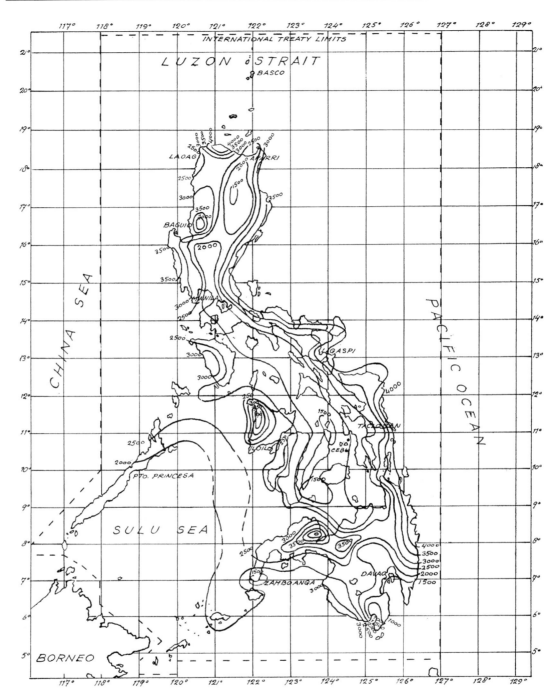

Fig.13. Distribution of mean annual rainfall (mm) in the Philippines.

stations have large rainfalls for almost all the months of the year. On the other hand, most of the western coastal stations do not derive an appreciable amount of rainfall from the Northeast Monsoon and the North Pacific Trades as shown by a relatively dry period during the Northeast Monsoon and North Pacific Trades seasons. The stations showing relatively small annual means of less than 2,000 mm are mostly located in valleys or plains or in places which are shielded from the dominant air streams by high mountain ranges.

182

Monthly and seasonal distribution of rainfall

Based on the monthly averages for all the 44 stations whose rainfall records were studied, August has the highest average with about 304 mm. April has the smallest average with about 100 mm. June to December may be considered the rainy months while the dry months cover the period from January to May. These generalizations for the rainy and dry months for the whole country are also true if each of the 3 regions were considered separately. It may be noted that the rainy months coincide with the typhoon season, the Southwest Monsoon and the first 3 months of the Northeast Monsoon season.

However, it is also evident that most of the eastern coastal stations have a marked rainy season during the period from October to March when the Northeast Monsoon is dominant. The western coastal stations have their rainy season during the period June–October at the height of the Southwest Monsoon and the tropical cyclone season.

Fig.14 is a graph showing the mean monthly rainfall distribution for ten selected stations in the Philippines. A further analysis of characteristics of rainfall at different stations in the Philippines will be presented in the section "Types of climates in the Philippines".

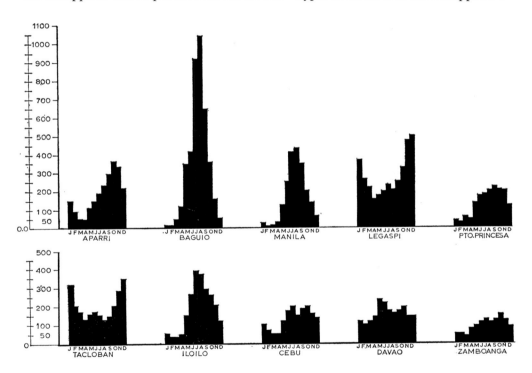

Fig.14. Mean monthly rainfall (mm) distribution, for ten stations in the Philippines.

Extreme rainfalls

Baguio, with an elevation of 1,482 m and located on the western slope of the Cordillera Central Ranges in northern Luzon, holds the annual, monthly, and 24 hour absolute maximum rainfall records for Luzon and the entire Philippines. The absolute maximum annual rainfall at Baguio was 9,038.3 mm in 1911, the absolute maximum monthly rainfall was 3,462 mm in August 1919, and the absolute maximum 24 hour rainfall was

TABLE VII

Station	Years	Jan.	Feb.	Mar.	Apr.	May	June	July	Aug.	Sep.	Oct.	Nov.	Dec.	Annual
Luzon														
Ambulong	9	6	3	3	5	11	15	19	19	18	15	13	11	11
Aparri	47	16	11	8	6	11	11	13	15	15	19	19	19	14
Aurora	7	12	6	8	5	9	13	19	17	18	19	15	14	13
Baguio	47	4	4	6	10	20	23	27	27	25	19	10	6	15
Baler	45	15	14	16	18	18	16	17	15	17	18	16	17	16
Basco	46	20	15	14	11	14	16	19	21	20	21	21	23	18
Cabanatuan	10	2	1	2	4	11	17	20	24	21	14	9	6	11
Casiguran	9	20	15	16	15	17	16	17	13	16	21	23	23	18
Daet	29	24	18	16	13	15	14	18	16	18	24	25	25	19
Dagupan	47	2	2	3	6	13	19	23	23	20	13	7	3	11
Iba	40	2	2	2	4	14	21	24	26	22	14	7	4	12
Infanta	23	25	20	18	17	19	18	19	18	19	24	25	27	21
Laoag	41	1	1	1	2	9	17	20	20	16	8	5	2	8
Legaspi	47	23	16	17	15	14	15	19	18	19	21	22	24	19
Lucena	28	17	9	9	6	10	14	16	15	16	19	19	20	14
Manila CO	65	6	3	4	4	12	17	24	23	22	19	14	11	13
Tuguegarao	46	7	5	5	6	11	12	15	15	14	14	14	11	11
Vigan	47	1	1	1	2	10	18	22	22	17	9	4	2	9
Virac	40	20	15	16	15	17	16	17	13	16	21	23	23	18
Manila MMO	9	5	3	3	3	8	15	19	23	22	17	12	10	12
Average for Luzon		11	8	8	8	13	16	19	19	19	17	15	14	14
Visayas														
Borongan	47	26	21	22	19	19	17	17	14	15	21	23	26	20
Calapan	31	18	13	12	11	13	16	18	15	16	20	21	22	16
Catarman	7	23	17	17	16	16	16	17	17	15	21	24	25	19
Catbalogan	34	19	15	16	14	15	17	18	16	18	21	21	21	18
Cebu	35	14	11	11	8	12	16	17	16	17	19	15	16	14
Coron	6	2	2	1	2	10	19	18	21	19	14	8	5	10
Cuyo	47	2	1	1	3	14	20	23	21	21	17	8	5	11
Dumaguete	38	14	9	9	6	11	14	15	14	15	17	15	16	13
Iloilo	47	11	7	7	6	14	18	21	20	19	18	15	14	14
Masbate	45	17	12	10	6	9	13	16	16	16	16	16	18	14
Pto. Princesa	9	4	3	4	5	13	15	15	17	17	17	15	9	11
Romblon	46	14	11	10	8	12	15	18	16	17	20	19	18	15
Roxas	47	17	11	9	8	14	19	20	17	17	21	20	19	16
Tacloban	46	22	17	18	15	16	17	17	15	16	19	21	23	18
Average for Visayas		14	11	10	9	13	17	18	17	17	19	17	17	15
Mindanao														
Cagayan de Oro	40	10	7	7	5	12	17	17	16	16	16	11	11	12
Cotabato	38	11	9	9	13	17	18	19	17	17	18	17	12	15
Davao	47	10	9	9	10	15	15	13	12	12	13	11	11	12
Dipolog	7	20	14	13	11	15	20	21	18	19	22	23	22	18
General Santos	7	8	7	7	9	13	15	14	15	12	12	12	11	11
Hinatuan	6	25	22	25	22	19	18	18	16	15	17	21	24	20
Jolo	52	9	8	9	11	16	17	15	14	15	17	17	14	13
Malaybalay	7	14	11	11	13	20	24	25	24	25	22	19	17	19
Surigao	47	24	20	21	18	14	13	15	13	14	18	21	25	18
Zamboanga	49	8	6	7	8	11	13	14	13	12	13	13	10	11
Average for Mindanao		10	8	8	9	11	12	12	11	11	12	12	11	11
Average for Philippines		13	10	10	9	14	16	18	18	17	18	16	15	15

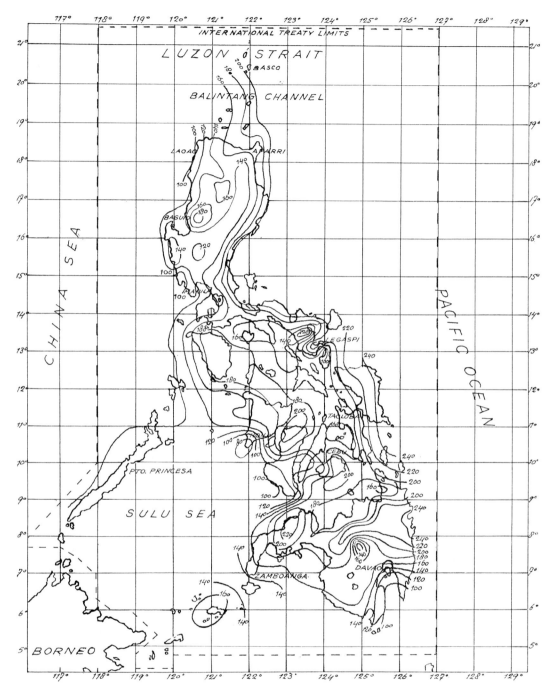

Fig.15. Distribution of mean annual number of rainy days in the Philippines.

1,168.1 mm on July 14–15, 1911. This 24 hour rainfall which was recorded from noon of July 14 to noon of July 15, 1911, during the passage of a typhoon across northern Luzon, was long recognized as a world's record for a period of 24 consecutive hours.[1] For the

---

[1] It has only recently been superseded by 1,870 mm of rain during a 24 hour period at Cilaos, Reunion Island, in the Indian Ocean. Typhoon Gloria also dumped 1,248 mm on Paishih, Taiwan, on September 10–11, 1963 (Editor-in-Chief).

Visayas, Borongan holds the absolute annual, monthly, and 24 hour maximum rainfall records with values of 6,789.9 mm, 2,193.3 mm and 571 mm, respectively. Similarly, for Mindanao, Surigao holds the annual, monthly and 24 hour absolute maximum rainfall records with values of 6,038.3 mm, 1,480.1 mm and 424.4 mm, respectively.

Number of rainy days

Closely associated with the amount of rainfall is the number of rainy days. The definition of rainy day as used here is a day having rainfall of 0.1 mm or more. The mean monthly and annual number of rainy days for 44 stations in the Philippines is shown in Table VII. It can be seen from Table VII that the mean annual number of rainy days is about 175. A large majority of the stations in the eastern coastal areas have mean annual number of rainy days more than the average for the entire Philippines while the majority of the stations in the western coastal areas have values less than the average for the entire Philippines. Again, the inland and sheltered stations have the lowest mean annual number of rainy days. Fig.15 shows the distribution of mean annual number of rainy days in the Philippines.

The months with the greatest number of rainy days are from June to December wherein the monthly value is greater than the monthly average of about 15 days. The months from January to May have monthly mean values of less than 15 days. These periods of monthly number of rainy days greater than 15 days coincide with the period of greatest rainfall amounts, and the period of monthly number of rainy days less than 15 days coincides with the period of small amounts of rainfall. Fig.16 is a graph showing the mean monthly distribution of the number of rainy days for ten selected stations in the Philippines.

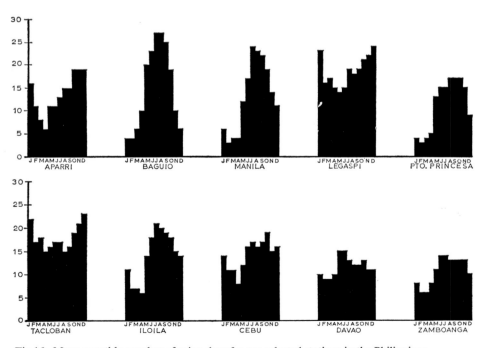

Fig.16. Mean monthly number of rainy days for ten selected stations in the Philippines.

Floods and droughts

A consequence of heavy rainfall, especially during the passage of tropical cyclones and the intensification of the monsoons, is that the big rivers in the islands, especially in Luzon and Mindanao, overflow their banks and may cause floods which persist from 1 to 5 days. These floods are almost yearly occurrences in big river basins.

During the dry months, on the other hand, continuous periods of rainless days (in excess of 30 consecutive days and sometimes exceeding 100 days), especially in the western coastal areas of Luzon, may result in droughts. However, unlike the floods, these droughts are not yearly occurrences but occur rather infrequently.

Humidity

Throughout the Philippines the humidity of the air is relatively high. This condition is mainly a result of extensive evaporation from the seas surrounding the islands, the rich vegetation, the moist air streams affecting the Philippines, and the large amount of rainfall. Table VIII shows the monthly and annual normal relative humidity for 44 stations in the Philippines.

The average annual relative humidity for the Philippines is about 82%. Annual means range from 76% for Cabanatuan and Cebu to 88% for Hinatuan. A large number of stations have annual means greater than the average value of 82%, while the rest have a smaller value. Although there is a generally good correlation between the mean annual amount of rainfall and the mean annual relative humidity, a number of the stations do not show this very well.

Most of the stations have high relative humidity during the period from June to November and low relative humidity during the period from December to May. Almost all the stations have monthly values of relative humidity greater than 70% and a large majority of the stations have more than 8 months of the year with relative humidity greater than 80%.

The diurnal behaviour of relative humidity in most stations is characterized by high values at night and early morning and low values during the day and early evening. However, variations from this general behaviour occur during rainfall during the day and early evening when the relative humidity values are high.

Cloudiness

This element is generally related to the humidity. However, cloudiness does not always show good correlation with rainfall because the cloudiness reported is a total of all the clouds present and a large portion of these clouds are not rain-producing clouds. This is supported by the fact that many of the stations with large mean annual cloudiness have small mean annual rainfall. Table IX shows the mean monthly and annual cloudiness for 44 stations in the Philippines and Fig.17 shows a graph of the monthly cloudiness for 10 selected stations in the Philippines.

The average annual cloudiness for 44 stations in the Philippines is about 6 oktas with values ranging from 4 oktas to 7 oktas. About half of the stations have average cloudiness of about 6 oktas or more while the rest have values of less than 6 oktas. For most of the

TABLE VIII

MEAN MONTHLY AND ANNUAL RELATIVE HUMIDITY FOR 44 STATIONS IN THE PHILIPPINES

Station	Jan.	Feb.	Mar.	Apr.	May	June	July	Aug.	Sep.	Oct.	Nov.	Dec.	Annual
Luzon													
Ambulong	78	75	71	71	75	81	84	84	86	84	81	81	79
Aparri	83	82	81	78	79	79	79	81	83	84	84	85	81
Aurora	82	81	78	76	74	74	77	77	77	77	79	81	78
Baguio	80	80	80	83	87	89	91	93	92	88	83	80	85
Baler	83	83	84	84	83	82	81	80	82	83	84	83	83
Basco	80	80	80	82	83	83	85	84	81	80	81	81	82
Cabanatuan	72	70	65	66	70	80	83	85	85	81	79	74	76
Casiguran	89	87	87	86	86	86	86	87	88	89	87	90	87
Daet	84	83	82	82	82	82	79	83	85	86	85	85	83
Dagupan	75	74	72	72	75	80	84	85	85	81	78	77	78
Iba	78	77	75	75	77	83	86	87	87	83	81	79	81
Infanta	81	78	76	73	78	81	83	83	84	84	84	83	81
Laoag	72	71	69	71	75	81	85	85	85	78	76	74	77
Legaspi	83	82	81	80	81	82	83	83	85	84	84	84	83
Lucena	84	84	80	79	78	81	81	82	83	84	83	84	82
Manila CO	77	73	70	69	74	80	83	84	84	83	81	80	78
Tuguegarao	81	77	74	70	70	75	78	79	81	83	85	85	78
Vigan	72	74	74	74	76	81	84	86	85	79	75	73	78
Virac	80	81	79	79	79	79	81	79	82	83	82	82	80
Manila MMO	78	73	68	66	70	79	82	85	86	84	82	81	78
Average for Luzon	80	78	76	76	78	81	83	80	84	83	82	81	80
Visayas													
Borongan	86	84	84	84	84	85	84	82	83	84	85	86	84
Calapan	84	81	78	77	78	82	83	84	84	84	84	84	82
Catarman	88	87	86	85	86	86	86	85	85	87	88	89	86
Catbalogan	83	83	81	81	81	82	81	80	82	84	85	85	82
Cebu	77	75	73	72	75	77	78	77	78	79	79	78	76
Coron	71	76	74	67	78	83	86	84	86	84	82	80	79
Cuyo	82	81	80	79	81	84	85	85	85	83	82	83	
Dumaguete	81	79	77	76	78	79	79	78	80	80	81	81	79
Iloilo	81	78	76	73	78	81	83	83	84	84	84	83	81
Masbate	84	82	80	78	78	80	82	83	83	83	84	85	82
Pto. Princesa	85	83	82	81	84	87	87	87	89	88	87	86	86
Romblon	82	81	80	72	78	79	82	83	84	84	83	82	81
Roxas	81	80	78	76	78	81	82	82	83	83	82	82	81
Tacloban	85	84	82	81	82	82	81	79	81	82	85	85	82
Average for Visayas	82	81	79	77	80	82	83	82	83	84	84	83	82
Mindanao													
Cagayan de Oro	80	79	77	75	77	80	80	79	80	81	81	82	79
Cotabato	80	79	77	79	82	83	84	84	84	83	83	82	82
Davao	82	79	78	79	82	84	83	83	83	82	83	83	82
Dipolog	84	81	80	79	82	84	84	83	84	85	85	85	83
General Santos	78	77	77	73	75	84	84	85	85	83	83	80	80
Hinatuan	90	89	89	89	88	88	87	85	87	87	88	89	88
Jolo	86	87	87	86	87	86	85	84	85	86	86	86	86
Malaybalay	82	81	80	78	82	85	86	86	86	85	85	84	83
Surigao	88	87	86	86	85	84	81	80	81	84	87	89	85
Zamboanga	83	82	81	83	84	85	85	85	85	85	86	84	84
Average for Mindanao	83	82	81	81	82	84	84	83	84	84	85	84	83
Average for Philippines	81	80	78	77	79	82	83	83	84	83	83	83	82

TABLE IX

MEAN MONTHLY AND ANNUAL CLOUDINESS (OKTAS) FOR 44 STATIONS IN THE PHILIPPINES

Station	Years	Jan.	Feb.	Mar.	Apr.	May	June	July	Aug.	Sep.	Oct.	Nov.	Dec.	Annual
Luzon														
Ambulong	14	5	5	4	3	5	6	6	6	6	6	6	5	5
Aparri	35	6	5	4	3	4	5	6	6	5	5	6	6	5
Aurora	7	6	5	5	4	5	5	6	6	6	6	6	6	5
Baguio	40	4	4	4	5	6	6	6	7	6	6	5	5	5
Baler	19	6	6	6	5	6	6	6	6	6	6	6	6	6
Basco	14	6	6	5	5	5	5	6	6	6	6	6	6	6
Cabanatuan	9	3	3	3	3	4	5	6	6	6	5	5	4	4
Casiguran	9	6	5	4	4	4	5	6	6	6	6	6	6	5
Daet	13	6	6	6	4	5	5	6	6	6	6	6	6	6
Dagupan	25	3	3	3	3	5	6	6	6	6	5	4	4	4
Iba	14	3	3	3	3	5	6	6	6	6	5	5	4	5
Infanta	9	6	6	6	6	6	6	6	6	6	7	7	7	6
Laoag	19	2	2	2	2	4	6	6	6	6	5	4	3	4
Legaspi	35	6	5	4	4	4	5	6	6	6	6	6	6	5
Lucena	9	6	5	5	5	6	6	7	6	6	6	6	6	6
Manila CO	59	5	4	4	3	5	6	6	6	6	6	6	5	5
Tuguegarao	16	6	6	5	4	5	6	6	6	6	6	6	6	6
Vigan	25	2	2	2	3	3	5	6	6	5	4	2	3	4
Virac	19	6	6	5	4	5	5	6	6	6	6	6	6	6
Manila MMO	8	6	5	4	4	5	6	6	7	7	6	6	6	6
Average for Luzon		5	5	4	4	5	6	6	6	6	6	5	5	5
Visayas														
Borongan	9	6	6	4	5	5	6	6	6	6	6	6	6	6
Calapan	7	7	7	6	5	6	7	7	7	7	7	7	7	7
Catarman	7	6	5	4	3	4	4	5	6	6	6	6	5	5
Catbalogan	19	5	5	5	4	5	6	6	6	6	6	6	6	5
Cebu	35	5	5	4	4	5	6	6	6	6	6	6	6	5
Coron	7	3	4	3	2	4	6	6	6	6	6	5	5	5
Cuyo	15	6	6	6	5	6	7	7	7	7	6	6	6	6
Dumaguete	14	6	6	6	5	6	6	6	6	6	6	6	6	6
Iloilo	35	5	5	4	4	6	6	6	6	6	6	6	5	5
Masbate	9	6	5	4	3	5	6	6	6	6	6	6	6	5
Pto. Princesa	7	5	5	4	5	6	6	6	6	6	6	6	6	6
Romblon	8	6	5	4	4	4	6	6	6	6	6	6	6	5
Roxas	14	6	6	5	4	5	6	6	6	6	6	6	6	6
Tacloban	20	6	6	5	5	6	6	6	6	6	6	6	6	6
Average for Visayas		6	5	5	4	5	6	6	6	6	6	6	6	6
Mindanao														
Cagayan de Oro	14	6	5	5	5	6	6	6	6	6	6	6	6	6
Cotabato	16	6	5	5	5	6	6	6	6	6	6	6	6	6
Davao	19	6	6	6	6	6	6	6	6	6	6	6	6	6
Dipolog	7	6	6	5	5	5	6	6	6	6	6	6	6	6
General Santos	7	3	5	3	4	4	5	4	4	4	3	4	4	4
Hinatuan	7	6	6	6	6	6	6	6	6	6	6	6	6	6
Jolo	19	6	6	6	6	6	6	6	6	6	6	6	6	6
Malaybalay	7	6	6	6	5	6	7	6	7	6	6	6	6	6
Surigao	35	6	6	6	5	5	6	6	6	6	6	6	6	6
Zamboanga	19	5	6	5	6	6	6	6	6	6	6	6	6	6
Average for Mindanao		6	6	5	5	6	6	6	6	6	6	6	6	6
Average for Philippines		5	5	5	4	5	6	6	6	6	6	6	6	6

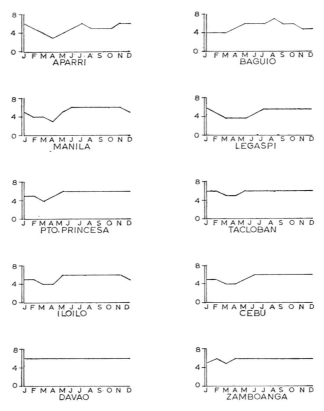

Fig.17. Mean monthly cloudiness (oktas) for ten selected stations in the Philippines.

stations, the cloudy months are June–December wherein the average cloudiness is about 6 oktas or more while the other months are less cloudy with average cloudiness of 5 oktas or less.

Surface winds

In the Philippines the winds are the usual composite of the major air currents, tropical cyclones and local circulations produced by diurnal and topographic effects.

Table X shows monthly and annual prevailing surface wind directions and mean speeds for 16 stations in the Philippines. From Table X it can be seen that although the prevailing wind direction at most stations conforms with the dominant air streams during the different months, in many of the stations the effects of local topography and diurnal effects produce prevailing winds which deviate somewhat from the winds that might be expected. Thus, during the time when the Northeast Monsoon is the dominant air stream, the prevailing winds at most of the stations are generally from the northeast quadrant, especially for the stations along the eastern coastal areas. During the Southwest Monsoon season, winds are generally from the southwest quadrant, especially for stations in the western coastal areas. On the other hand, the other air streams affecting the Philippines are not very evident. Only a few stations on the eastern coastal areas show clearly the presence of the North Pacific Trades, while no station clearly shows the predominance of the South Pacific Trades or the Temperate Zone Westerlies in any month.

TABLE X

MEAN MONTHLY AND ANNUAL PREVAILING WINDS AND AVERAGE WIND SPEEDS (m/sec) FOR 16 STATIONS IN THE PHILIPPINES

Station	Jan.	Feb.	Mar.	Apr.	May	June	July	Aug.	Sep.	Oct.	Nov.	Dec.	Annual average
Manila CO	NE/2	E/2	SE/2	SE/3	SE/3	SW/3	SW/3	SW/4	SW/3	NE/2	NE/2	NE/2	SW/3
Baguio	SE/4	SE/4	SE/3	SE/3	SE/4	SE/3	W/5	W/5	W/4	SE/3	SE/4	SE/3	SE/4
Pto. Princesa	NE/1	NE/1	NE/1	W/1	W/1	W/1	W/1	W/1	W/1	W/1	W/1	NE/1	W/1
Manila MMO	E/3	SE/4	SE/4	SE/4	SE/4	SE/3	SW/3	SW/3	SE/2	SE/2	SE/2	E/2	SE/3
Laoag	N/3	N/3	N/3	NW/3	W/3	S/3	SSW/3	SSW/3	SSW/2	N/3	N/4	N/3	N/3
Zamboanga	NE/2	N/2	W/2	N/2	N/2	N/2	N/2	W/2	W/2	W/2	W/2	W/2	W/2
Jolo	NE/3	NE/3	NE/3	S/2	S/3	S/3	S/4	S/3	S/3	S/3	S/3	NE/3	S/3
Iloilo	NE/6	NE/5	NE/6	NE/5	NE/3	SW/3	SW/4	SW/4	SW/3	NE/3	NE/4	NE/5	NE/4
Cebu	NE/4	NE/3	NE/4	NE/3	NE/3	SW/3	SW/3	SW/4	SW/3	SW/3	NE/3	NE/3	NE/3
Dumaguete	N/2	NE/2	NE/2	NE/1	NE/1	NW/1	SW/1	SSW/1	SSW/1	NE/1	NW/1	NE/2	NE/1
Cagayan de Oro	N/2	N/2	N/2	N/2	NE/2	NE/2	N/2	N/2	N/2	N/2	N/2	N/2	N/2
Aparri	NE/4	NE/3	NE/4	NE/3	NE/3	S/3	S/3	S/3	S/3	NE/4	NE/4	NE/4	NE/3
Legaspi	NE/4	NE/3	NE/3	NE/3	NE/2	SW/2	SW/3	SW/3	SW/3	NE/3	NE/3	NE/4	NE/3
Tacloban	NW/3	NW/3	NW/3	NW/3	SE/3	S/3	W/2	W/2	W/2	W/3	W/3	NW/3	NW/3
Davao	N/3	N/3	N/2	N/2	N/2	S/2	S/2	S/2	S/2	N/2	N/2	N/2	N/2
Surigao	NE/3	NE/3	NE/3	NE/2	E/2	SW/2	SW/3	SW/3	SW/3	SW/2	NE/2	NE/2	NE/2
Prevailing average	NE/3	NE/3	NE/3	NE/3	NE/2	SW/2	SW/3	SW/3	SW/2	NE/2	NE/3	NE/3	NE/3

The annual average wind speed for 16 stations in the Philippines is rather low, being only 3 m/sec. Of course the wind speed is highly variable, especially during the passage of tropical cyclones through or near the Philippines. The wind speeds at each station during the presence of a tropical cyclone in the vicinity of the Philippines generally depends on their distances from the cyclone and also on local topographic effects. There has not been an accurate measurement of the maximum winds occurring at the stations in the direct path of tropical cyclones due to the fact that most of the anemometers in use in the Philippines are destroyed when the wind speed exceeds 50 m/sec. However, it may be assumed that wind speeds during direct passage of tropical cyclones over a place, especially on the exposed eastern coastal areas, occasionally exceed 50 m/sec and possibly exceed 75 m/sec. In Manila, the absolute maximum wind speed recorded was about 53 m/sec on 20 October 1882. The wind speed may also increase considerably above the average value during surges or intensification of the monsoons; up to values exceeding 15 m/sec in some instances.

Diurnal changes are usually observed at most stations. The effects of land and sea breezes at coastal stations, and mountain and valley breezes for inland stations are usually apparent. The wind speed has the usual maximum during the day and the common nocturnal minimum at most stations.

Atmospheric pressure

Although this is not generally considered to be an important climatic element in the tropics, its behaviour in the Philippines will nevertheless be presented for the sake of completeness. Table XI shows the mean monthly and annual mean sea level pressures

TABLE XI

MEAN MONTHLY AND ANNUAL MEAN SEA LEVEL PRESSURES (MBAR) FOR 44 STATIONS IN THE PHILIPPINES

Station	Jan.	Feb.	Mar.	Apr.	May	June	July	Aug.	Sep.	Oct.	Nov.	Dec.	Annual
Luzon													
Ambulong	1012.7	1012.3	1011.7	1010.1	1008.6	1008.3	1007.4	1007.4	1008.1	1008.9	1009.9	1011.1	1009.7
Aparri	1015.2	1014.3	1013.0	1010.8	1008.7	1007.3	1006.0	1005.6	1007.3	1009.9	1012.3	1013.9	1010.4
Aurora	1013.2	1012.4	1012.5	1011.1	1009.8	1009.3	1009.1	1008.4	1009.2	1009.3	1009.8	1011.6	1010.5
Baguio	1011.6	1010.5	1010.2	1009.4	1007.8	1006.7	1006.8	1005.9	1005.7	1007.0	1007.3	1010.2	1008.3
Baler	1015.1	1013.8	1013.9	1012.1	1009.5	1008.8	1008.6	1007.4	1009.0	1010.0	1011.0	1013.8	1011.1
Basco	1016.0	1015.0	1013.5	1011.6	1008.8	1007.2	1005.7	1005.3	1007.4	1010.3	1013.2	1015.2	1010.8
Cabanatuan	1014.2	1013.0	1013.0	1011.5	1010.2	1009.3	1009.4	1008.6	1009.6	1010.2	1010.8	1013.2	1011.1
Casiguran	1014.6	1013.2	1013.4	1011.6	1009.9	1008.2	1007.8	1006.7	1008.5	1009.7	1010.5	1013.4	1010.6
Daet	1013.1	1012.6	1012.3	1010.8	1009.2	1008.3	1007.2	1006.9	1007.9	1009.1	1010.1	1011.7	1009.9
Dagupan	1012.4	1012.0	1011.2	1009.7	1008.3	1007.5	1006.5	1006.5	1007.3	1008.5	1009.7	1010.9	1009.2
Iba	1013.0	1012.5	1012.1	1010.8	1009.6	1008.7	1008.9	1008.3	1009.2	1009.6	1010.1	1011.8	1010.4
Infanta	1014.9	1013.9	1013.9	1012.6	1010.5	1009.3	1009.1	1008.0	1009.4	1010.2	1011.1	1013.5	1011.4
Laoag	1012.7	1012.1	1011.3	1009.7	1008.2	1007.1	1005.9	1005.9	1007.0	1008.6	1010.0	1012.5	1009.2
Legaspi	1012.0	1012.1	1011.6	1010.2	1008.7	1008.2	1007.2	1007.0	1007.6	1008.7	1009.2	1010.3	1009.4
Lucena	1014.3	1013.6	1013.8	1012.4	1011.1	1010.0	1009.6	1008.9	1009.7	1010.3	1010.4	1013.0	1011.1
Manila CO	1012.7	1012.6	1011.8	1010.1	1008.8	1008.3	1005.8	1005.5	1006.9	1009.4	1011.6	1013.3	1009.9
Tuguegarao	1014.5	1013.5	1012.2	1010.5	1008.5	1007.3	1006.3	1006.2	1007.2	1008.7	1009.9	1011.1	1009.3
Vigan	1012.6	1012.2	1011.5	1009.9	1008.4	1007.5	1006.3	1006.2	1007.2	1008.7	1009.9	1011.1	1009.3
Virac	1013.3	1012.7	1012.8	1011.7	1010.4	1009.5	1009.3	1008.3	1009.4	1009.1	1010.0	1011.8	1010.7
Manila MMO	1013.4	1012.5	1012.4	1010.9	1009.6	1008.9	1009.0	1008.3	1009.2	1009.6	1010.3	1013.0	1010.6
Average for Luzon[1]	1013.7	1013.0	1012.5	1010.9	1009.3	1008.4	1007.7	1007.2	1008.3	1009.4	1010.5	1012.4	1010.3
Visayas													
Borongan	1012.0	1011.4	1011.8	1010.8	1009.7	1009.2	1009.2	1008.4	1009.1	1009.1	1009.1	1010.5	1010.0
Calapan	1014.1	1013.1	1013.1	1011.6	1010.2	1009.3	1009.3	1008.5	1009.4	1009.9	1010.5	1012.7	1011.0
Catarman	1012.6	1012.9	1012.2	1011.1	1009.8	1009.3	1009.2	1008.4	1009.1	1009.1	1009.3	1010.8	1010.3
Catbalogan	1012.1	1011.4	1011.6	1010.7	1009.5	1009.1	1009.3	1008.6	1009.2	1009.2	1009.0	1009.3	1009.9
Cebu	1010.7	1010.9	1010.6	1009.6	1008.8	1008.6	1008.1	1008.1	1008.5	1008.8	1008.7	1009.4	1009.2
Coron	1012.3	1011.6	1011.9	1011.1	1010.1	1009.9	1007.5	1009.7	1009.9	1010.0	1009.8	1011.1	1010.4
Cuyo	1011.1	1010.7	1010.9	1010.0	1009.1	1009.0	1009.2	1008.7	1009.5	1009.4	1009.1	1010.2	1009.7
Dumaguete	1010.6	1010.0	1010.2	1009.5	1008.7	1008.7	1008.9	1008.6	1008.9	1008.8	1008.5	1009.5	1009.2
Iloilo	1010.6	1010.7	1010.2	1009.2	1008.5	1008.4	1007.9	1008.1	1008.4	1008.7	1008.7	1007.1	1008.9
Masbate	1012.7	1011.8	1011.8	1010.8	1009.5	1009.0	1009.3	1008.7	1009.0	1009.1	1009.1	1010.8	1010.1
Pto. Princesa	1009.7	1010.8	1011.0	1010.2	1009.4	1008.8	1009.6	1009.4	1009.7	1009.6	1009.5	1010.4	1009.8
Romblon	1012.8	1012.1	1012.8	1011.4	1010.2	1009.5	1009.4	1008.9	1009.6	1009.9	1010.2	1011.7	1010.7
Roxas	1011.6	1011.7	1011.1	1009.8	1008.7	1008.4	1007.7	1007.7	1008.1	1008.8	1009.1	1010.1	1009.4
Tacloban	1011.1	1011.3	1011.1	1010.0	1007.8	1008.6	1007.9	1007.9	1008.2	1008.8	1009.0	1009.8	1009.2
Average for Visayas	1011.7	1011.4	1011.4	1010.4	1009.3	1009.0	1008.7	1008.5	1009.0	1009.2	1009.2	1010.2	1009.8
Mindanao													
Cagayan de Oro	1008.7	1010.1	1010.4	1009.7	1009.2	1009.4	1009.8	1009.6	1010.0	1009.7	1009.0	1009.8	1009.6
Cotabato	1009.8	1009.6	1009.8	1009.4	1009.4	1009.5	1010.1	1009.9	1010.2	1009.7	1008.8	1009.2	1009.6
Davao	1011.0	1010.7	1010.9	1010.4	1009.8	1010.0	1010.2	1009.9	1010.4	1010.0	1009.2	1009.8	1010.2
Dipolog	1010.6	1010.1	1010.5	1009.7	1009.2	1009.2	1009.6	1009.3	1009.7	1009.6	1009.0	1009.7	1009.7
General Santos	1010.5	1010.1	1010.3	1010.0	1010.0	1010.1	1010.6	1010.5	1010.7	1010.3	1009.5	1010.0	1010.2
Hinatuan	1009.3	1008.8	1009.2	1008.5	1007.9	1007.9	1008.0	1007.7	1008.0	1007.9	1007.3	1008.2	1008.2
Jolo	1009.1	1008.6	1009.0	1008.6	1008.2	1008.5	1009.8	1008.9	1009.0	1008.9	1008.3	1008.7	1008.8
Malaybalay	1006.6	1006.8	1007.2	1006.3	1006.1	1006.2	1006.8	1006.5	1006.6	1006.5	1005.6	1006.5	1006.5
Surigao	1010.4	1010.6	1010.4	1009.6	1008.7	1008.6	1008.0	1008.0	1008.3	1008.5	1008.7	1009.2	1009.1
Zamboanga	1009.2	1009.4	1009.2	1008.6	1008.6	1008.8	1008.7	1008.9	1009.2	1009.0	1008.5	1008.5	1008.9
Average for Mindanao[2]	1009.8	1009.8	1010.0	1009.4	1009.0	1009.1	1009.4	1009.2	1009.5	1009.3	1008.7	1009.2	1009.4
Average for Philippines[3]	1012.2	1011.7	1011.6	1010.4	1009.2	1008.7	1008.4	1008.1	1008.1	1008.8	1009.3	1009.7	1009.9

[1] Baguio not included

[2] Malaybalay not included

[3] Baguio and Malaybalay not included

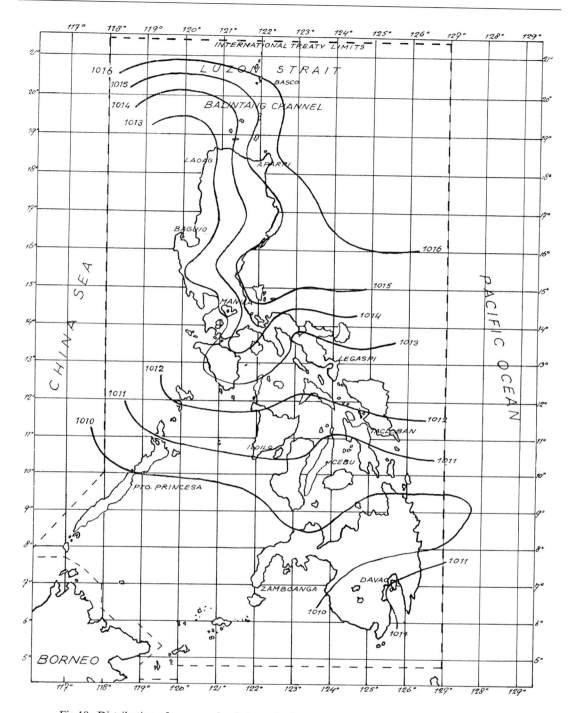

Fig.18. Distribution of mean sea level atmospheric pressure (mbar) in the Philippines for January.

for 44 stations in the Philippines. The average annual pressure for 42 stations in the Philippines (the highly elevated stations of Baguio and Malaybalay are excluded) is 1,009.9 mbar. Values range from 1,008.2 mbar to 1,011.4 mbar.

The effects of the extreme seasons, winter and summer, are generally evident with highest pressures occurring in the winter months, lowest pressures in the summer months, and

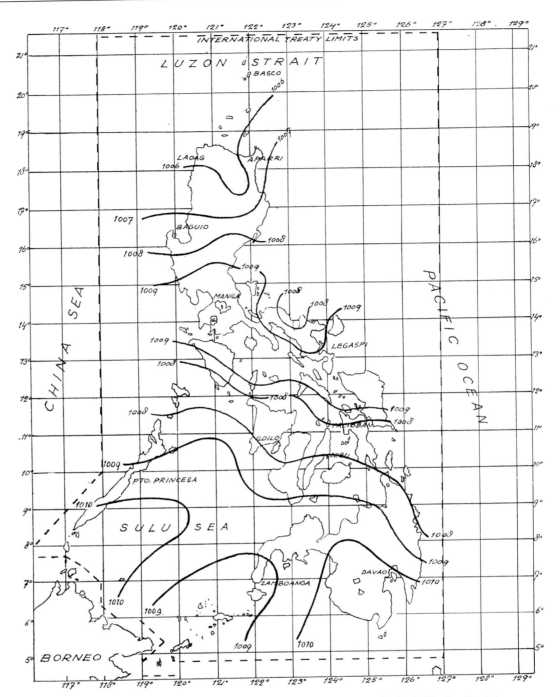

Fig.19. Distribution of mean sea level atmospheric pressure (mbar) in the Philippines for July.

intermediate values during the autumn and spring months. Of course the low pressures in summer in the Philippines are partly caused by the effects of tropical cyclones. The pressures generally decrease equatorward in winter, increase equatorward in summer, and remain almost the same throughout the Philippines during spring and autumn. Fig.18–19 show the distribution of mean sea level pressures in the Philippines for January

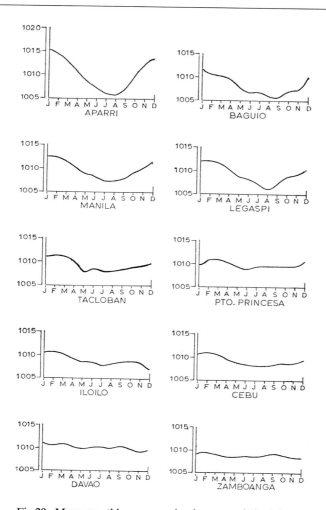

Fig.20. Mean monthly mean sea level pressure (mbar) for ten selected stations in the Philippines.

and July, respectively. Fig.20 is a graph showing the monthly mean sea level pressure distribution for ten selected stations in the Philippines.

The ordinary diurnal pressure variations for most of the stations in the Philippines follow the characteristic diurnal variation for tropical stations, with two maxima at about 10 a.m. and 10 p.m. and two minima at about 4 a.m. and 4 p.m. Differences between adjacent maximum and minimum values generally range from 2.5 to 4.5 mbar.

With respect to extreme values, the maximum values recorded in the different stations do not vary much from year to year, but the minimum values vary considerably due to the effects of tropical cyclones. It has not been possible to determine the absolute minimum pressure recorded in the Philippines because of the loss of some records, but some of the lowest values recorded during tropical cyclones are as follows: 920.1 mbar at San Policarpio Bay, Samar Island, on 25 September 1905; 924.1 mbar at Tacloban on 24 November 1912; and 925.2 mbar at Tanauan, Leyte, on 12 October 1897. On the other hand, maximum pressure values occasionally exceed 1,025 mbar during the winter months.

Types of climates of the Philippines

There are a number of classifications of climates. Most familiar are those by Köppen and Thornthwaite. These two systems of classification are most commonly used because they are based on universally available data and have specific numerical values for defining the boundaries of climatic groups and types. The classifications of climates of the Philippines presented here are based on: Köppen's classifications, 1930; CORONAS' classification, 1920; and a modification of Thornthwaite's classification by HERNANDEZ, 1954.

Climates of the Philippines classified according to Köppen

This system is based on annual and monthly means of temperature and precipitation using vegetation limits in the selection of climatic boundaries. In this system, five climatic types are applicable to the Philippines which are described as follows:

Af climates

Tropical wet climate; temperature of coolest month above 18°C; rainfall of the driest month is at least 60 mm. Within this climate there is a minimum of seasonal variation in temperature and precipitation. Both remain high during the year. Areas in the Philippines belonging to this type are: the greater portion of Mindanao; Samar; Leyte; Bohol; Catanduanes; western Mindoro; and the eastern coastal areas of central Luzon.

Aw climates

Tropical wet and dry climate; temperature of the coolest month above 18°C; distinct dry season in low-sun period or winter; characterized by a marked seasonal rhythm of rainfall; at least one month must have less than 60 mm of rainfall. Places in the Philippines with this type of climate are: the valleys and plains of Luzon; Cebu; and small areas along the coasts of northern and southwestern Mindanao.

Am climates

Tropical monsoon climate; temperature of the coolest month above 18°C; short dry season, but with total rainfall so great that the ground remains sufficiently wet throughout the year to support rainforests; rainfall of the driest month is below 60 mm. It is intermediate between Af and Aw, resembling Af in amount of rainfall and Aw in seasonal distribution. Areas in the Philippines belonging to this type of climate are: the western and northeastern coastal areas of Luzon; western Mindoro; Panay; the western coastal areas of Negros; most of Palawan; Masbate; and southwestern Mindanao.

Cf climates

Warm temperate rainy climate; average temperature of coldest month below 18°C but above −3°C; average temperature of warmest month over 10°C; no distinct dry season, the driest month of summer receives more than 30 mm of rainfall. Some of the mountain

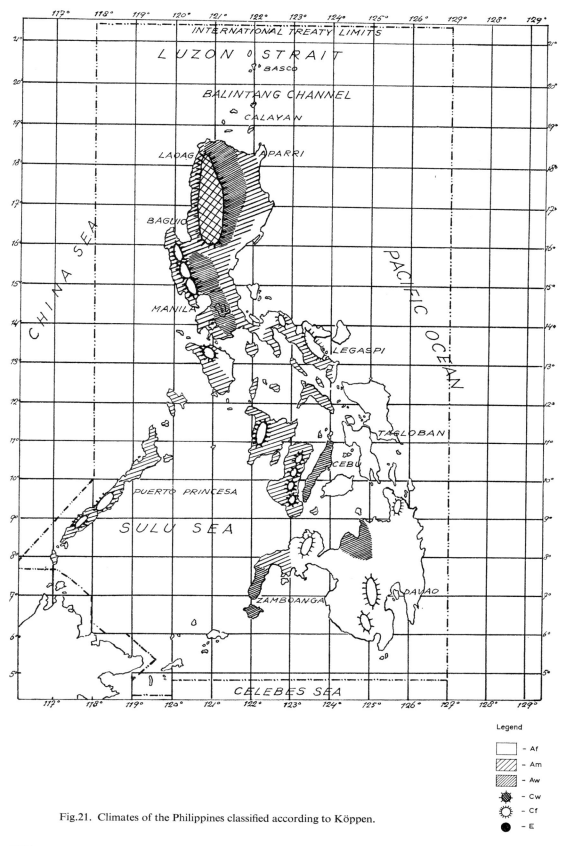

Fig.21. Climates of the Philippines classified according to Köppen.

areas of Mindanao, Negros, Panay, Palawan, Mindoro and southern Luzon belong to this type of climate.

Cw climates

Warm temperate rainy climate; average temperature of coldest month below 18°C but above −3°C; average temperature of warmest month over 10°C; winter dry; at least ten times as much rain in the wettest month of summer as in the driest month of winter. Only the high mountain ranges of northwestern Luzon belong to this type of climate. Fig.21 shows the climates of the Philippines classified according to Köppen.

Climates of the Philippines classified according to Coronas

Recognizing the fact that temperature differences in the Philippines are very small while rainfall variations are large, CORONAS (1920) devised a system of climatic classification for the Philippines based solely on rainfall characteristics. Using the average monthly distribution of rainfall and taking into consideration the greater or less prevalence of either of the two most important rain periods, he divided this monthly distribution of rainfall into four types of which two are altogether opposite types, and two others are intermediates.

First type

Two pronounced seasons, dry in winter and spring, wet in summer and autumn. Only the cyclonic or summer rainfall prevails, the other being hardly noticeable. Hence, the dry season of winter lasts from three to 6 or 7 months. Strictly speaking, a dry month should be understood as a month with less than 50 mm of rain. Sometimes, however, a month with more than even 100 mm of rain is considered a dry month, especially if it comes after three or more very dry months. This type of climate is found in all the stations on the western part of the islands of Luzon; Mindoro; Negros; Palawan; and the western and southern part of Panay.

Second type

No dry season, with a very pronounced maximum rain period in winter. There is in the regions of this type much cyclonic or summer and autumn rainfall, but the maximum monthly rainfall is generally that of December and January, while the monthly amounts of rain for the summer and autumn months are far from being so great. There is not a single dry month in regions of this type. The minimum monthly rainfall occurs in some places in spring, and in other places in summer. The regions having this type of climate are most of southeastern Luzon; practically the whole of Samar; eastern Leyte; and a great portion of eastern Mindanao.

Third or intermediate A type

No very pronounced maximum rain period, with a short dry season lasting only from one to three months. This type is intermediate between the preceeding two, although it

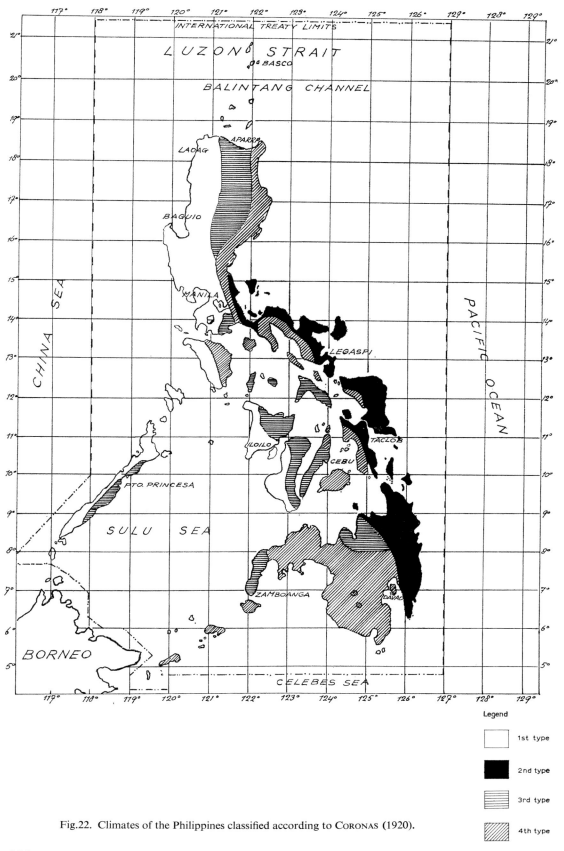

Fig.22. Climates of the Philippines classified according to Coronas (1920).

Legend

1st type

2nd type

3rd type

4th type

resembles the first type more closely since it has a short dry season. The short dry season experienced in regions of this type occurs in some places in winter, and in other places in spring. Places with this type of climate include the Cagayan Valley of northern Luzon; northeastern Panay; eastern Negros; central and southern Cebu; part of northern and southwestern Mindanao; and a large portion of eastern Palawan.

Fourth or intermediate B type

No very pronounced maximum rain period and no dry season. This also is an intermediate type between the first and the second, but it resembles the second more closely since it has no dry season. Both cyclonic and Northeast Monsoon rains, as well as thunderstorm rains, are experienced in these regions with not a single dry month during the year. The minimum monthly rainfall occurs in spring, although in some places it takes place in January. Areas with this type of climate are the Batanes Islands; the easternmost part of northern and central Luzon; a portion of southern Luzon; eastern Mindoro; Marinduque; a small portion of Samar; western Leyte; northern Cebu; the small islands of Bohol, Jolo and Basilan; and a large portion of central and western Mindanao.

Fig.22 shows the areas covered by these four types of climates under Coronas' system of classification.

Climates of the Philippines classified according to Hernandez

This system is based on ratios of dry months to wet months. It is an outgrowth of the work of Thornthwaite, Mohr, and Schmidt. By determining the number of dry and wet months, year by year, using the precipitation limits proposed by Mohr, and taking the average values of the result, HERNANDEZ (1954) believed that the smoothing effect caused by simple averaging would be prevented. A month with rainfall of less than 60 mm is considered a dry month while a month with rainfall of more than 100 mm is considered a wet month.

To determine the type of climate, the quotient Q of the total number of dry months divided by the total number of wet months were computed. The border lines between the various types lie at $Q = 1.5a/(12 - 1.5a)$ where a, in the case of the Philippines, is an integer from 0 to 6. The types of climates under this system[1], with their characteristics and distribution in the Philippines, are as follows:

Type A

Wet; rainy throughout the year with at most $1\frac{1}{2}$ dry months; Q less than 0.14. This type of climate is found in most of the eastern coastal areas of Luzon; Samar; eastern sections of Masbate; Leyte; and a large portion of eastern and central Mindanao.

Type B

Humid; rain well or evenly distributed throughout the year with at most 3 dry months; Q equals 0.14 or more but less than 0.33. Portions of northern and southern Luzon;

---

[1] To quickly determine a type of climate under this classification, see Fig.1, p. 225 of this volume (Editor).

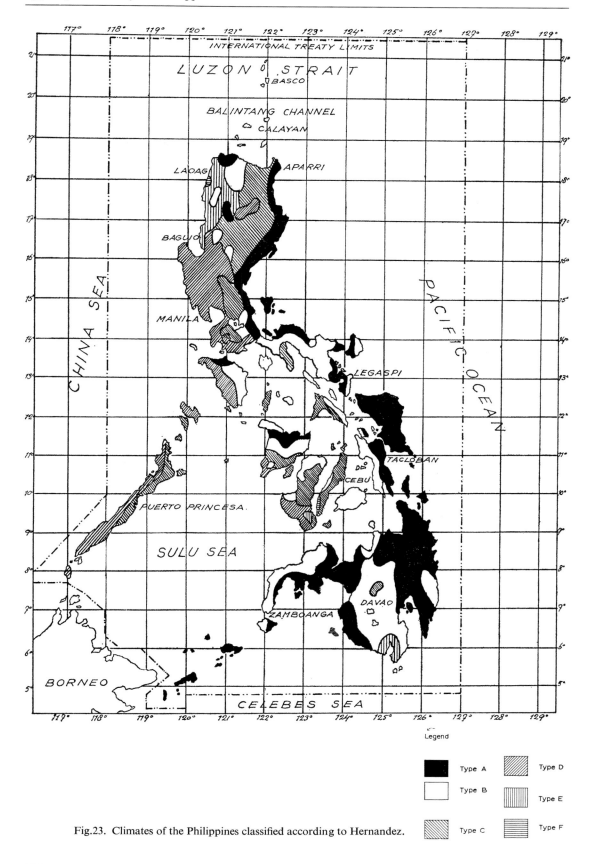

Fig.23. Climates of the Philippines classified according to Hernandez.

Legend

Type A   Type D

Type B   Type E

Type C   Type F

eastern Mindoro; Panay; northern Negros; western Leyte; Bohol; portions of western Cebu; and portions of eastern, central and western Mindanao have this type of climate.

Type C

Moist; rain sufficiently distributed with at most $4\frac{1}{2}$ dry months; Q equals 0.33 or more, but less than 0.60. Regions with this type of climate include a large portion of the valleys and plains of northern and central Luzon; western Masbate; eastern Palawan; southern Panay; central Negros; eastern Cebu; and a small area in northern Mindanao.

Type D

Dry; rain not sufficiently distributed with at most 6 dry months; Q equals 0.60 or more, but less than 1.00. Places with this type of climate are found in southwestern Luzon; western Mindoro; western Palawan; and portions of southwestern Negros and Panay.

Type E

Arid; there are more dry than wet months; at most there are $4\frac{1}{2}$ wet months; Q equals 1.00 or more but less than 1.67. Areas having this type of climate include northwestern Luzon and a small portion of southern Mindanao.

Type F

Barren; deficient rainfall with less than 3 wet months; Q equals 1.67 or more but less than 3.00. Only a very small area in northwestern Luzon has this type of climate.

Fig.23 shows the distribution of climatic types in the Philippines under this system of classification.

Acknowledgements

The authors wish to acknowledge the valuable help given by Messrs. S. R. Zarate, D. A. Liwag, A. S. Santos, P. dela Cruz, and B. S. Lomotan, of the Philippine Weather Bureau, in the gathering, processing, and analysis of data and in the preparation of illustrations and tables. The authors are also indebted to Messrs. J. J. Tecson, J. F. Lirios, C. P. Arafiles, C. C. Reyes and R. E. Ledesma, also of the Philippine Weather Bureau, for reading the manuscript and for their constructive criticisms and suggestions.

References

ALGUE, J., 1904. *The Cyclones of the Far East*. Bureau of Printing, Manila, 283 pp.

BLAIR, T. A., 1943. *Climatology*. Prentice-Hall, New York, N.Y., 484 pp.

CHIN, P. C., 1958. *Tropical Cyclones in the Western Pacific and China Sea Area from 1884 to 1953*. Royal Observatory, Hong Kong, 94 pp.

CORONAS, J., 1920. *The Climate and Weather of the Philippines, 1903 to 1918*. Bureau of Printing, Manila, 195 pp.

DEPPERMAN, C. E., 1934. *The Upper Air at Manila.* Bureau of Printing, Manila, 29 pp.

DEPPERMAN, C. E., 1936. *Outline of Philippine Frontology.* Bureau of Printing, Manila, 27 pp.

DEPPERMAN, C. E., 1943. *Upper Air Circulation over the Philippines.* Bureau of Printing, Manila, 64 pp.

ESTOQUE, M. A., 1952. *Circulation and Convergence over the Tropical Northwestern Pacific.* Weather Bureau, Manila, 31 pp.

HAURWITZ, B. and AUSTIN, J. M., 1944. *Climatology.* McGraw-Hill, New York, N.Y., 410 pp.

HERNANDEZ, P. P., 1954. *Rainfall Types Based on Ratios of Dry Months to Wet Months.* Weather Bureau, Manila, 32 pp.

RAMAGE, C. C., 1959. *Notes on the Meteorology of the Tropical Pacific and Southeast Asia.* Meteorol. Div., Hawaii Inst. Geophys., Univ. Hawaii, Honolulu, 144 pp.

RIEHL, H., 1954. *Tropical Meteorology.* McGraw-Hill, New York, N.Y., 392 pp.

SELGA, M., REPPETTI, W. C. and ADAMS, W., 1931. *Oceanographic Papers*, Vol. III (1–10). Bureau of Printing, Manila, 210 pp.

SVERDRUP, H. U., 1943. *Oceanography for Meteorologists.* Prentice-Hall, New York, N.Y., 235 pp.

TREWARTHA, G. T., 1954. *An Introduction to Climate.* McGraw-Hill, New York, 402 pp.

WIEDERANDERS, C. J., 1961. *Analysis of Monthly Mean Resultant Winds for Standard Pressure Levels over the Pacific.* Meteorol. Div., Hawaii Inst. Geophys., Univ. Hawaii, Honolulu, 83 pp.

Data sources

AIR MINISTRY METEOROLOGICAL OFFICE, 1944. *Notes on the Weather of the Far East.* H.M. Stationery Office, London, 12 pp.

HEADQUARTERS ARMY AIR FORCES, 1944. *Interpretation of Weather Over the Pacific.* U.S. Govt. Printing Office, Washington, D.C., 23 pp.

RESEARCH SECTION, U.S. WEATHER FLEET CENTRAL, HEADQUARTERS 5TH AIR FORCE, 1944. *Clouds and Weather Over the Philippines.* 914 th Eng. Air Force Headquarters Company, 13 pp.

U.S. DEPARTMENT OF COMMERCE AND HYDROGRAPHIC OFFICE, U.S. NAVY DEPARTMENT, 1961. *Climatological and Oceanographic Atlas for Mariners.* U.S. Govt. Printing Office, Washington D.C. 125 charts.

U.S. NAVY DEPARTMENT, 1941. *Climatology of Asiatic Stations* U.S. Govt. Printing Office, Washington, D.C., 99pp.

WEATHER BUREAU, PHILIPPINES, 1961. *Climate of the Different Ports of Entry in the Philippines.* Weather Bureau, Manila, 32 pp.

WEATHER BUREAU, PHILIPPINES, 1960. *Monthly Average Rainfall and Rainy Days in the Philippines.* Weather Bureau, Manila, 15 pp.

WEATHER BUREAU, PHILIPPINES, 1960. *Monthly Average Relative Humidity in the Philippines.* Weather Bureau, Manila, 4 pp.

WEATHER BUREAU, PHILIPPINES, 1960. *Monthly Average Temperature in the Philippines.* Weather Bureau, Manila, 3 pp.

TABLE XII

CLIMATIC TABLE FOR APARRI
Latitude 18°22′N, longitude 121°38′E, elevation 4 m

Month	Mean pressure (mbar) at mean sea level	Mean daily temp. (°C)	Mean daily temp. range (°C)	Temp. extremes (°C)		R.H. (%)	Mean precip. (mm)	Max. precip. (24 h)
				highest	lowest			
Jan.	1015.2	23.2	17.8	33.4	15.6	83	144.0	119.1
Feb.	1014.3	23.8	19.1	33.9	14.8	82	89.7	72.4
Mar.	1013.0	25.3	20.3	36.1	15.8	81	54.6	92.0
Apr.	1010.8	27.1	22.4	38.4	16.0	78	48.5	137.2
May	1008.7	28.1	17.9	37.9	20.0	79	111.0	131.3
June	1007.3	28.5	17.8	38.9	21.1	79	144.0	167.9
July	1006.0	28.1	16.1	37.1	21.0	79	172.5	235.5
Aug.	1005.6	27.9	17.0	37.0	20.0	81	234.2	806.2
Sept.	1007.3	27.5	15.0	36.0	21.0	83	295.1	252.5
Oct.	1009.9	26.5	16.0	35.0	19.0	84	366.8	236.5
Nov.	1012.3	25.2	17.7	34.0	16.3	84	335.8	310.1
Dec.	1013.9	23.9	18.2	33.2	15.0	85	217.9	180.6
Annual	1010.3	26.3	24.1	38.9	14.8	81	2259.6	806.2

Month	Mean evapor. (mm)	Number of days			Mean cloudiness (okta)	Mean hours sunshine	Wind	
		precip.	thunder-storms	fog			most freq. direction	mean speed (m/sec)
Jan.		16	0		6		NE	4
Feb.		11	0		5		NE	3
Mar.		8	1		4		NE	4
Apr.		6	7		3		NE	3
May		11	9		4		NE	3
June		11	15		5		S	3
July		13	12		6		S	3
Aug.		15	12		6		S	3
Sept.		15	7		5		S	3
Oct.		19	6		5		NE	4
Nov.		19	1		6		NE	4
Dec.		19	0		6		NE	4
Annual		163	70		5		NE	3

TABLE XIII

CLIMATIC TABLE FOR BAGUIO
Latitude 16°25′N, longitude 120°36′E, elevation 1482 m

Month	Mean pressure (mbar) at mean sea level	Mean daily temp. (°C)	Mean daily temp. range (°C)	Temp. extremes (°C)		R.H. (%)	Mean precip. (mm)	Max. precip. (24 h)
				highest	lowest			
Jan.	1011.6	16.9	22.4	29.7	7.3	80	19.6	92.0
Feb.	1010.5	17.4	21.0	28.3	7.3	80	19.1	48.3
Mar.	1010.2	18.3	19.4	29.7	10.3	80	48.3	70.1
Apr.	1009.4	18.9	19.3	29.3	10.0	83	121.7	129.0
May	1007.8	19.2	15.6	28.4	12.8	87	351.8	415.8
June	1006.7	18.9	16.8	28.6	11.8	89	422.1	496.6
July	1006.8	18.4	14.3	26.8	12.5	91	921.5	879.8
Aug.	1005.9	18.0	14.2	27.0	12.8	93	1042.9	642.9
Sept.	1005.7	18.2	14.0	26.9	12.9	92	649.7	799.9
Oct.	1007.0	18.3	16.0	27.3	11.3	88	361.7	689.6
Nov.	1007.3	18.1	18.6	27.8	9.2	83	162.6	407.4
Dec.	1010.2	17.5	20.6	28.2	7.6	80	55.9	183.9
Annual	1008.3	18.2	22.4	29.7	7.3	85	4176.9	879.8

Month	Mean evapor. (mm)	Number of days			Mean cloudiness (okta)	Mean hours sunshine	Wind	
		precip.	thunder-storms	fog			most freq. direction	mean speed (m(sec)
Jan.		4	0		4		SE	4
Feb.		4	0		4		SE	4
Mar.		6	1		4		SE	3
Apr.		10	6		5		SE	3
May		20	10		6		SE	4
June		23	9		6		SE	3
July		27	8		6		W	5
Aug.		27	6		7		W	5
Sept.		25	5		6		W	4
Oct.		19	4		6		SE	3
Nov.		10	0		5		SE	4
Dec.		6	0		5		SE	3
Annual		181	49		5		SE	4

TABLE XIV

CLIMATIC TABLE FOR MANILA CENTRAL OFFICE
Latitude 14°35'N, longitude 120°59'E, elevation 16 m

Month	Mean pressure (mbar) at mean sea level	Mean daily temp. (°C)	Mean daily temp. range (°C)	Temp. extremes (°C)		R.H. (%)	Mean precip. (mm)	Max. precip. (24 h)
				highest	lowest			
Jan.	1012.7	25.0	20.7	35.2	14.5	77	22.9	136.2
Feb.	1012.6	25.5	20.0	35.6	15.6	73	10.9	43.7
Mar.	1011.8	26.8	20.5	36.7	16.2	70	16.8	59.9
Apr.	1010.1	28.3	20.8	38.0	17.2	69	32.3	143.0
May	1008.8	28.6	18.6	38.6	20.0	75	128.3	218.2
June	1007.3	27.9	16.0	37.6	21.6	80	252.5	252.7
July	1007.4	27.1	15.5	36.3	20.8	84	413.5	293.6
Aug.	1007.4	27.0	14.6	35.2	20.6	84	436.9	323.8
Sept.	1007.9	26.9	14.5	35.3	20.8	85	353.1	336.0
Oct.	1009.2	26.7	15.6	35.1	19.5	83	195.3	194.3
Nov.	1010.1	25.9	17.6	34.4	16.8	82	137.9	278.4
Dec.	1011.5	25.2	18.9	34.6	15.7	80	68.3	128.3
Annual	1009.8	26.7	24.1	38.6	14.5	78	2068.7	336.0

Month	Mean evapor. (mm)	Number of days			Mean cloudiness (okta)	Mean hours sunshine	Wind	
		precip.	thunder-storms	fog			most freq. direction	mean speed (m/sec)
Jan.		6	0		5		NE	2
Feb.		3	0		4		E	2
Mar.		4	0		4		SE	2
Apr.		4	2		3		SE	3
May		12	7		5		SE	3
June		17	8		6		SW	3
July		24	9		6		SW	3
Aug.		23	8		6		SW	4
Sept.		22	8		6		SW	3
Oct.		19	6		6		NE	2
Nov.		14	2		6		NE	2
Dec.		11	1		5		NE	2
Annual		159	51		5		SW	3

TABLE XV

CLIMATIC TABLE FOR LEGASPI
Latitude 13°08′N, longitude 123°44′E, elevation 19 m

Month	Mean pressure (mbar) at mean sea level	Mean daily temp. (°C)	Mean daily temp. range (°C)	Temp. extremes (°C)		R.H. (%)	Mean precip. (mm)	Max. precip. (24 h)
				highest	lowest			
Jan.	1012.0	25.7	15.5	32.7	17.2	83	366.0	239.0
Feb.	1012.1	25.9	17.0	33.7	16.7	82	264.9	185.2
Mar.	1011.6	26.7	16.9	33.9	17.0	81	218.2	154.2
Apr.	1010.2	27.7	17.2	36.1	18.9	80	158.0	193.0
May	1008.7	28.2	17.3	37.2	19.9	81	178.3	202.7
June	1008.2	28.1	15.6	37.0	21.4	82	194.1	378.0
July	1007.2	27.4	15.8	36.6	20.8	83	235.0	224.3
Aug.	1007.0	27.4	15.8	36.9	21.1	83	209.3	142.5
Sept.	1007.6	27.3	15.6	36.0	20.4	85	251.7	216.2
Oct.	1008.7	27.1	16.9	35.3	18.4	84	313.4	263.9
Nov.	1009.2	26.7	15.8	34.4	18.6	84	478.8	297.4
Dec.	1010.3	26.2	15.6	33.1	17.5	84	502.9	248.4
Annual	1009.4	27.0	20.5	37.2	16.7	83	3370.6	378.0

Month	Mean evapor. (mm)	Number of days			Mean cloudiness (okta)	Mean hours sunshine	Wind	
		precip.	thunder-storms	fog			most freq. direction	mean speed (m/sec)
Jan.		23	0		6		NE	4
Feb.		16	0		5		NE	3
Mar.		17	1		4		NE	3
Apr.		15	3		4		NE	3
May		14	7		4		NE	2
June		15	7		5		SW	2
July		19	5		6		SW	3
Aug.		18	7		6		SW	3
Sept.		19	6		6		SW	3
Oct.		21	3		6		NE	3
Nov.		22	3		6		NE	3
Dec.		24	0		6		NE	4
Annual		223	42		5		NE	3

TABLE XVI

CLIMATIC TABLE FOR TACLOBAN
Latitude 11°15′N, longitude 125°00′E, elevation 21 m

Month	Mean pressure (mbar) at mean sea level	Mean daily temp. (°C)	Mean daily temp. range (°C)	Temp. extremes (°C)		R.H. (%)	Mean precip. (mm)	Max. precip. (24 h)
				highest	lowest			
Jan.	1011.1	25.7	15.9	34.7	18.8	85	318.8	246.6
Feb.	1011.3	25.8	17.2	34.8	17.6	84	209.0	132.6
Mar.	1011.1	26.4	17.9	35.9	18.0	82	177.8	178.6
Apr.	1010.0	27.3	17.8	38.0	20.2	81	133.9	136.1
May	1007.8	27.8	17.3	37.9	20.6	82	163.3	325.9
June	1008.6	27.6	14.9	36.2	21.3	82	172.0	244.1
July	1007.9	27.5	16.3	37.8	21.5	81	160.5	244.4
Aug.	1007.9	27.7	17.4	38.0	20.6	79	133.1	88.4
Sept.	1008.2	27.6	15.6	37.2	21.6	81	150.9	116.1
Oct.	1008.8	27.2	16.2	36.0	19.8	82	207.3	150.1
Nov.	1009.0	26.7	15.4	35.2	19.8	85	288.3	206.5
Dec.	1009.8	26.0	16.5	34.0	17.5	85	353.3	192.8
Annual	1009.2	26.9	20.5	38.0	17.5	82	2468.2	325.9

Month	Mean evapor. (mm)	Number of days			Mean cloudiness (okta)	Mean hours sunshine	Wind	
		precip.	thunder-storms	fog			most freq. direction	mean speed (m/sec)
Jan.		22	1		6		NW	3
Feb.		17	0		6		NW	3
Mar.		18	2		5		NW	3
Apr.		15	4		5		NW	3
May		16	11		6		SE	3
June		17	10		6		S	3
July		17	9		6		W	2
Aug.		15	11		6		W	2
Sept.		16	12		6		W	2
Oct.		19	9		6		W	3
Nov.		21	4		6		W	3
Dec.		23	2		6		NW	3
Annual		216	75		6		NW	3

TABLE XVII

CLIMATIC TABLE FOR ILOILO
Latitude 10°42′N, longitude 122°34′E, elevation 14 m

Month	Mean pressure (mbar) at mean sea level	Mean daily temp. (°C)	Mean daily temp. range (°C)	Temp. extremes (°C)		R.H. (%)	Mean precip. (mm)	Max. precip. (24 h)
				highest	lowest			
Jan.	1010.6	25.8	14.7	33.2	18.5	81	59.2	118.6
Feb.	1010.7	26.1	17.4	35.4	18.0	78	38.1	79.5
Mar.	1010.2	26.9	16.0	35.3	19.3	76	35.6	79.2
Apr.	1009.2	28.2	15.8	37.0	21.2	73	51.6	85.1
May	1008.5	28.3	15.1	36.8	21.7	78	153.4	223.8
June	1008.4	27.7	14.3	35.7	21.4	81	264.7	154.9
July	1007.9	27.2	15.4	35.2	19.8	83	389.6	233.4
Aug.	1008.1	26.9	14.7	34.8	20.1	83	370.3	222.2
Sept.	1008.4	27.1	15.9	36.0	20.1	84	293.6	151.9
Oct.	1008.7	26.9	13.9	34.7	20.8	84	262.9	182.9
Nov.	1008.7	26.6	14.5	34.8	20.3	84	206.5	237.5
Dec.	1007.1	26.1	15.5	33.8	18.3	83	120.9	172.2
Annual	1008.9	27.0	19.0	37.0	18.0	81	2246.4	237.5

Month	Mean evapor. (mm)	Number of days			Mean cloudiness (okta)	Mean hours sunshine	Wind	
		precip.	thunder-storms	fog			most freq. direction	mean speed (m/sec)
Jan.		11	0		5		NE	6
Feb.		7	1		5		NE	5
Mar.		7	1		4		NE	6
Apr.		6	5		4		NE	5
May		14	10		6		NE	3
June		18	10		6		SW	3
July		21	9		6		SW	4
Aug.		20	7		6		SW	4
Sept.		19	8		6		SW	3
Oct.		18	7		6		NE	3
Nov.		15	3		6		NE	4
Dec.		14	2		5		NE	5
Annual		170	63		5		NE	4

TABLE XVIII

CLIMATIC TABLE FOR CEBU
Latitude 10°20′N, longitude 123°54′E, elevation 42 m

Month	Mean pressure (mbar) at mean sea level	Mean daily temp. (°C)	Mean daily temp. range (°C)	Temp. extremes (°C)		R.H. (%)	Mean precip. (mm)	Max. precip. (24 h)
				highest	lowest			
Jan.	1010.7	26.3	14.9	33.1	18.2	77	104.9	180.8
Feb.	1010.9	26.4	14.3	32.6	18.3	75	71.1	122.9
Mar.	1010.6	27.2	15.3	34.4	19.1	73	54.6	83.8
Apr.	1009.6	28.2	13.7	34.7	21.0	72	52.8	125.5
May	1008.8	28.4	14.0	35.4	21.4	75	120.9	111.1
June	1008.6	28.0	14.7	35.0	20.3	77	177.0	147.3
July	1008.1	27.6	13.3	34.6	21.3	78	196.6	170.2
Aug.	1008.1	27.7	14.2	35.2	21.0	77	152.7	80.3
Sept.	1008.5	27.6	13.5	34.4	20.9	78	186.9	128.5
Oct.	1008.8	27.5	13.5	34.2	20.7	79	201.4	299.7
Nov.	1008.7	27.1	15.1	34.4	19.3	79	162.8	100.6
Dec.	1009.4	26.7	14.1	33.4	19.3	78	137.7	321.6
Annual	1009.2	27.4	17.2	35.4	18.2	76	1619.4	321.6

Month	Mean evapor. (mm)	Number of days			Mean cloudiness (okta)	Mean hours sunshine	Wind	
		precip.	thunder-storms	fog			most freq. direction	mean speed (m/sec)
Jan.		14	0		5		NE	4
Feb.		11	0		5		NE	3
Mar.		11	0		4		NE	4
Apr.		8	1		4		NE	3
May		12	6		5		NE	3
June		16	10		6		SW	3
July		17	8		6		SW	3
Aug.		16	9		6		SW	4
Sept.		17	11		6		SW	3
Oct.		19	8		6		SW	3
Nov.		15	2		6		NE	3
Dec.		16	1		6		NE	3
Annual		172	56		5		NE	3

TABLE XIX

CLIMATIC TABLE FOR PUERTO PRINCESA
Latitude 09°45′N, longitude 118°44′E, elevation 16 m

Month	Mean pressure (mbar) at mean sea level	Mean daily temp. (°C)	Mean daily temp. range (°C)	Temp. extremes (°C)		R.H. (%)	Mean precip. (mm)	Max. precip. (24 h)
				highest	lowest			
Jan.	1009.7	26.9	12.8	33.4	20.6	85	37.8	47.8
Feb.	1010.8	27.2	13.4	33.4	20.0	83	28.2	57.4
Mar.	1011.0	27.7	13.8	34.9	21.1	82	60.2	116.3
Apr.	1010.2	28.5	12.8	35.0	22.2	81	42.2	50.3
May	1009.4	28.5	12.2	34.9	22.7	84	135.6	121.7
June	1008.8	27.4	13.3	35.0	21.7	87	170.7	56.9
July	1009.6	27.2	11.8	33.4	21.6	87	181.1	78.2
Aug.	1009.4	27.1	11.6	32.8	21.2	87	201.4	81.0
Sept.	1009.7	27.1	11.6	33.3	21.7	88	220.7	60.7
Oct.	1009.6	27.2	11.5	33.2	21.7	88	206.8	134.1
Nov.	1009.5	27.2	11.8	32.9	21.1	87	204.2	202.4
Dec.	1010.4	26.9	13.4	34.0	20.6	86	118.4	90.4
Annual	1009.8	27.4	15.0	35.0	20.0	86	1607.3	134.1

Month	Mean evapor. (mm)	Number of days			Mean cloudiness (okta)	Mean hours sunshine	Wind	
		precip.	thunder-storms	fog			most freq. direction	mean speed (m/sec)
Jan.		4	0		5		NE	1
Feb.		3	0		5		NE	1
Mar.		4	5		4		NE	1
Apr.		5	3		5		W	1
May		13	3		6		W	1
June		15	0		6		W	1
July		15	0		6		W	1
Aug.		17	0		6		W	1
Sept.		17	0		6		W	1
Oct.		17	0		6		W	1
Nov.		15	1		6		W	1
Dec.		9	0		6		NE	1
Annual		134	12		6		W	1

TABLE XX

CLIMATIC TABLE FOR DAVAO
Latitude 07°04′N, longitude 125°36′E, elevation 20 m

Month	Mean pressure (mbar) at mean sea level	Mean daily temp. (°C)	Mean daily temp. range (°C)	Temp. extremes (°C) highest	lowest	R.H. (%)	Mean precip. (mm)	Max. precip. (24 h)
Jan.	1011.0	26.3	17.7	34.7	17.0	82	118.1	75.2
Feb.	1010.7	26.6	18.7	36.7	18.0	79	103.4	87.1
Mar.	1010.9	27.2	19.3	36.7	17.4	78	119.1	118.1
Apr.	1010.4	27.7	17.5	36.6	19.1	79	142.0	101.1
May	1009.8	27.6	17.1	37.3	20.2	82	235.7	131.6
June	1010.0	27.0	14.5	35.2	20.7	84	216.7	99.1
July	1010.2	26.8	15.2	35.2	20.0	83	175.5	94.7
Aug.	1009.9	26.9	17.5	36.0	18.5	83	161.5	94.0
Sept.	1010.4	27.0	14.5	34.5	20.0	83	175.5	123.7
Oct.	1010.0	27.1	16.2	35.4	19.2	82	192.5	133.4
Nov.	1009.2	27.0	17.1	36.2	19.1	83	143.3	88.6
Dec.	1009.8	26.6	18.5	34.7	16.2	83	145.0	153.7
Annual	1010.2	27.0	21.1	37.3	16.2	82	1928.3	153.7

Month	Mean evapor. (mm)	Number of days precip.	thunder-storms	fog	Mean cloudiness (okta)	Mean hours sunshine	Wind most freq. direction	mean speed (m/sec)
Jan.		10	1		6		N	3
Feb.		9	2		6		N	3
Mar.		9	3		6		N	2
Apr.		10	4		6		N	2
May		15	8		6		N	2
June		15	6		6		S	2
July		13	7		6		S	2
Aug.		12	5		6		S	2
Sept.		12	7		6		S	2
Oct.		13	11		6		N	2
Nov.		11	4		6		N	2
Dec.		11	1		6		N	2
Annual		140	59		6		N	2

TABLE XXI

Latitude 06°54′N, longitude 122°04′E, elevation 6 m

Month	Mean pressure (mbar) at mean sea level	Mean daily temp. (°C)	Mean daily temp. range (°C)	Temp. extremes (°C)		R.H. (%)	Mean precip. (mm)	Max. precip. (24 h)
				highest	lowest			
Jan.	1009.2	26.3	18.3	34.9	16.6	83	51.8	128.0
Feb.	1009.4	26.4	19.1	34.7	15.6	82	52.6	156.5
Mar.	1009.2	26.8	18.2	35.7	17.5	81	38.4	74.7
Apr.	1008.6	27.1	16.0	35.7	19.7	83	52.6	61.0
May	1008.6	27.1	14.4	35.1	20.7	84	89.7	55.4
June	1008.8	26.8	14.4	34.8	20.4	85	112.8	56.1
July	1008.7	26.6	15.0	35.0	20.0	85	126.0	92.7
Aug.	1008.9	26.8	16.1	35.1	19.0	85	113.0	136.6
Sept.	1009.2	26.7	14.9	34.8	19.9	85	122.9	116.8
Oct.	1009.0	26.6	16.4	35.8	19.4	85	155.4	121.4
Nov.	1008.5	26.6	16.7	35.2	18.5	86	120.7	163.3
Dec.	1008.5	26.4	17.8	34.5	16.7	84	88.1	161.0
Annual	1008.9	26.7	20.2	35.8	15.6	84	1124.0	163.3

Month	Mean evapor. (mm)	Number of days			Mean cloudiness (okta)	Mean hours sunshine	Wind	
		precip.	thunder-storms	fog			most freq. direction	mean speed (m/sec)
Jan.		8	1		5		NE	2
Feb.		6	1		6		N	2
Mar.		7	1		5		W	2
Apr.		8	5		6		N	2
May		11	6		6		N	2
June		13	3		6		N	2
July		14	2		6		N	2
Aug.		13	1		6		W	2
Sept.		12	3		6		W	2
Oct.		13	1		6		W	2
Nov.		13	1		6		W	2
Dec.		10	1		6		W	2
Annual		128	26		6		W	2

Chapter 4

Climate of Indonesia

M. SUKANTO

Introduction

Indonesia is situated between the mainland of Asia and Australia and is bordered by the Pacific and Indonesian (Indian) Oceans. Two-thirds of the total area of Indonesia is water whereas the country itself, with its thousands of islands spread out along the equator, is rich in volcanoes and mountain ranges.

Due to the structure of the islands and their geographical location, the air over Indonesia is warm and humid. The climate is almost entirely controlled by the monsoon, especially in the southern part of the country. The Northern Hemisphere winter corresponds to the wet season whereas the dry season occurs during the Southern Hemisphere winter. The region in which air coming from the Northern Hemisphere meets air from the Southern Hemisphere plays an important role in the weather of Indonesia. This region is known as the Intertropical Zone of Convergence. Along this region the weather is usually bad, and sometimes a line of cumulonimbus clouds with tops reaching 15 km or more can be found.

In general, the wet season prevails in the area north of the Convergence Zone, and the dry season prevails south of it. Although the season is called "dry", this does not necessarily mean that there is no rain at all. A wet and dry season are distinguishable but, in general, there is no month without rainfall. In fact, according to Köppen's classification, a tropical rainforest climate predominates in Indonesia.

The most unstable air occurs along the doldrums of low pressure which coincide with the Intertropical Convergence Zone. This zone follows the seasonal shift of the sun, but how precisely this occurs is still unknown. The direction of wind and the precipitation phenomena are the most usual indicators of this region where the trade winds from both hemispheres meet. Sometimes, however, this region disappears without any trace and suddenly appears again far behind or ahead of its original position. It reaches its southernmost position in January, not exceeding latitude 15°S. This means that all parts of the country have a wet season.

It must be kept in mind that in addition to this important Zone of Convergence, the water content of the air, together with orographical lifting, may cause heavy rain showers for long periods. Diurnal heating over land areas may also produce convective showers.

These are the reasons why there is an appreciable amount of rainfall in the dry season. The same reasoning applies to the frequency of thunderstorms, which is probably the highest in the world (HAURWITZ and AUSTIN, 1944). It is a pity that the information on the number of thunderstorms in Indonesia is obtained by visual methods and not by recording instruments, which are obviously more accurate.

It must also be pointed out, that the data presented in the tables accompanying this paper are far from perfect and should be improved in the future.

Temperature

Table I shows that the monthly mean temperature does not change very much. Roughly speaking, it is almost constant throughout the year. The deviation from month to month is small; much smaller, in fact, than the daily temperature range. The horizontal contrast of the temperature is very small. On the other hand, the variation in the vertical direction is much more important. In this connection, the altitude above sea level could take over the role of the temperature as far as climate is concerned.

TABLE I

MEAN TEMPERATURE (°C), 1931–1960

Station	Number of years	J	F	M	A	M	J	J	A	S	O	N	D	Year max.	min.	mean
Ambon	14	27.6	27.9	27.6	27.0	26.0	25.0	25.2	25.5	26.0	26.8	27.3	27.7	30.3	23.1	26.7
Bandjarmasin	8	26.2	26.9	26.7	26.4	26.6	26.8	26.2	26.4	27.2	27.7	27.0	26.7	33.0	22.8	26.8
Balikpapan	8	26.3	26.2	26.3	26.5	26.9	26.6	26.2	26.5	26.8	26.8	26.6	26.8	30.0	24.0	26.5
Djakarta	28	26.2	26.3	26.8	27.2	27.3	27.0	26.7	27.0	27.4	27.4	27.0	26.6	31.2	23.7	26.9
Medan	22	26.4	26.9	27.3	27.7	28.1	27.8	27.4	27.2	26.9	26.9	26.5	26.3	31.8	22.6	27.1
Menado	16	24.9	24.8	25.2	25.8	25.8	25.8	25.9	26.2	25.7	25.9	25.5	25.0	30.3	22.3	25.6
Padang	12	26.5	27.0	27.1	21.1	29.6	29.4	29.3	29.1	31.0	28.8	28.3	28.5	30.4	24.9	29.0
Palembang	10	23.9	26.0	26.4	27.1	27.4	26.8	26.4	26.3	26.9	26.9	26.4	26.0	31.7	22.9	26.5
Kupang	20	27.0	26.8	26.7	26.9	26.8	26.2	25.6	25.9	24.0	25.0	28.1	27.4	33.1	21.8	27.0
Surabaja	20	26.6	26.4	26.8	26.9	26.6	26.6	26.0	26.1	26.9	27.8	28.0	26.8	31.1	23.5	26.8

Irian Barat (West New Guinea) with Sukarno Peak (Carstenztop) is an interesting subject for research workers in the field of climatology. There are various types of climate in this part of the country ranging from tropical-rainy over the coastal region to eternal snow which covers Sukarno Peak and some other peaks in the central range of this island. This area might be considered a natural laboratory for this branch of science.

It has been suggested by BRAAK (1929) that between sea level and 1500 m the decrease in the mean temperature with height is 0.6°C/100 m. From this point upward this figure becomes smaller and takes on rate of 0.55°C/100 m.

According to the last expedition carried out jointly by Japan and Indonesia, the snowline is situated at about 4,300 m. When this party reached the top of Sukarno Peak (5,030 m) on 1 March 1964 at 01:30 p.m., the temperature was 6°C. It is a pity that there was no time to record temperatures at other hours. It was reported that in the early morning the peak was free of fog. About 07:00 a.m. the fog started to develop and hung there the whole day until a few hours before sunrise. By using this information together with the frequency distribution of fog with respect to temperature as put forward by PETTERSSEN (1958), one might estimate the temperature at about − 5°C during the morning hours.

The following formula for the island of Djawa (Java) concerning the mean temperature change with altitude was also introduced by BRAAK (1929):

$$t = 26.3 - 0.61\,h$$

whereby t and h stand respectively for the mean temperature in °C and the altitude of the station above sea level as expressed in hectometers. This formula gives the mean temperature for various stations which differ only less than one degree from the actual mean.

Hydrometeors

Rainfall is the most important hydrometeor in Indonesia. In general, the amount of rainfall is more than 2,000 mm/year.

As usual, heavy rains occur along the Intertropical Zone of Convergence. Since this zone

TABLE II

MEAN RAINFALL (MM), 1931–1960

Station	Height (m)	Number of years	J	F	M	A	M	J	J	A	S	O	N	D	Total
Djakarta	7	30	334	241	201	141	113	97	61	52	78	91	155	196	1760
Surabaja	5	15	329	278	292	209	118	90	41	18	12	43	151	256	1837
Padang	2	30	361	252	355	409	340	289	250	350	459	573	581	545	4764
Medan Polonia	14	15	165	103	118	163	215	132	174	207	207	256	268	166	2174
Palembang	14	11	255	265	309	285	155	128	102	86	85	202	343	265	2480
Bandjarmasin	20	28	302	264	315	214	160	141	104	112	138	131	233	279	2393
Balikpapan	3	14	186	170	248	219	220	265	258	246	195	154	194	242	2597
Menado	4	13	393	400	247	293	254	256	246	170	204	218	294	377	3352
Kupang	2	11	391	263	292	60	41	12	7	0	1	9	56	165	1297
Ambon	1	30	136	119	129	256	559	522	571	503	249	170	198	145	3457
Bosnik (Biak)	5	11	243	217	266	219	226	231	239	219	233	167	198	189	2647
Merauke	30	11	311	265	362	321	185	161	130	160	149	150	171	224	2589

moves up and down, crossing the equator in a northerly as well as southerly direction, there are places which show two maxima in the course of the year. This is best illustrated in Table II for the station Medan Polonia on Sumatra where the two extreme values occur in May and November. The same phenomenon is also demonstrated at Padang and Palembang on Sumatra, and on the island Biak. Only the time of occurrence is different due to the shift of the convergence zone.

One should also consider other factors, as mentioned above, which also play a role in the formation of precipitation. Afternoon showers, for instance, with an intensity of 0.5 mm/min caused by local intensive heating, are very common. Also, the flow of air over the mountain ranges in Sumatra, Djawa and Irian Barat might be accompanied by appreciable precipitation. Such processes may also occur during the dry season.

Generally speaking, the southeastern part of the archipelago has less annual rainfall than other parts of the country. It is a matter of the rain produced along the path of the air streaming from west to east during the wet season. The water content simply becomes smaller and smaller as the air streams from west to east. In the dry season, the air comes from the dry air of Australia.

Rainfall at Ambon, on the island Ambon, is different from other stations. The maximum occurs in July, as can be seen in Table II. This does not mean that it occurs in the dry season. The draining process of the water content along the path of the wind plays a role in the wet monsoon. On the other hand, the dry air from Australia becomes moist on its way over the Banda Sea. In this manner the deviation of its rainfall from the other stations might be explained.

The amount of precipitation also varies with altitude. In the first instance, rainfall will increase, reach a maximum and afterwards decrease with altitude. This phenomenon is strongly related to moisture and adiabatic cooling when the air streams up the mountainous regions. It could be understood that by such a process the air will become drier.

Gaps in the mountain ranges may reduce the amount of rainfall since part of the air will pass through the gaps and adiabatic cooling will not occur.

Both factors will determine the location of maximum rainfall with respect to altitude. This tendency can be seen on the island of Djawa, but it is still difficult to determine points of maximum precipitation due to a shortage of observations.

The last expedition to Sukarno Peak (December 1963–March 1964) did not encounter such phenomena. Rainfall observations were made by this party up to 3,560 m. At that altitude the daily rainfall was 23.8 mm whereas at 1,740 m, a value of only 11.1 mm was recorded. According to the report made at 4,050 m, there were no days without rain or showers; the weather was very bad. On the snowfield, a snowfall of about 20 cm was observed by the climbing party during one night.

According to present records (BERLAGE, 1949), the maximum annual rainfall is 7,069 mm. This was observed at Baturaden, central Djawa, which has an altitude of 700 m. The minimum value is 574 mm and occurs in the Palu Valley of Sulawesi (Celebes). The altitude of this valley is almost at sea level. One would expect that the true maximum might occur over the central range in Irian Barat. Since there are no continous records for this region however, rainfall data cannot be presented here.

Another interesting hydrometeor is the formation of frost. This phenomena may be found in the mountainous regions where humidity is low and where nocturnal radiation is well established. The first condition is necessary for cooling below the freezing point. If the formation of dew were to take place earlier, it would mean that the cooling process would be obstructed by the liberated heat of condensation. Both conditions could be met during a real dry season.

According to BRAAK (1929), frost occurs in the highlands of Pangalengan (1,500 m) on west Djawa during July, August and September. When freezing occurs, the temperature range and the relative humidity range are very large, for instance, 24.2°C and 60% on 24 August 1923. Frost has also been observed on the Dieng Plateau (2,100 m) in east Djawa, Pangrango (3,023 m) in west Djawa, Lalidjiwo (2,500 m) and the Yang Plateau (2,180 m) in central Djawa, Kalisat (1,100 m) in east Djawa and, of course, in the central range of Irian Barat.

Fog may develop over swampy land, large rivers with particularly steep banks and valleys or plateaus in the mountain regions. Humidity and nocturnal radiation are the leading factors which produce fog. Tropical, weak land winds with clear nights during the dry monsoon fulfil the requirements for nocturnal radiation processes. Air trapped in valleys or plateaus makes the cooling process better, especially when the air has been moistened by evaporation due to intensive solar radiation over water surface or by rain which usually occurs in the afternoon.

Palembang in Sumatra and Bandjarmasin in Kalimantan (Borneo) are the best examples of morning fogs which develop because of nocturnal radiation during the dry monsoon. In the plateau of Bandung in west Djawa, fog formation is also caused by nocturnal radiation.

Monsoon

Table III demonstrates that pressure distribution over Indonesia does not show a sharp contrast. The distance between the isobars, therefore, is very large. Consequently, the wind is usually weak, being in most cases about 5 m/sec.

TABLE III

MEAN AIR PRESSURE (MBAR), 1931–1960

Station	Number of years	J	F	M	A	M	J	J	A	S	O	N	D	Mean
Ambon	14	1006.7	1006.9	1007.0	1007.0	1007.6	1008.2	1008.5	1009.0	1009.0	1008.3	1007.0	1006.4	1007.8
Bandjarmasin	8	1009.5	1009.7	1009.2	1008.6	1008.3	1009.2	1009.5	1009.8	1009.8	1009.3	1008.0	1009.4	1009.2
Balikpapan	8	1008.3	1007.7	1007.9	1008.4	1008.2	1009.0	1009.1	1008.9	1009.5	1008.9	1008.3	1009.8	1008.6
Djakarta	28	1009.9	1009.9	1009.1	1009.6	1009.6	1009.7	1010.4	1010.4	1010.7	1010.5	1009.7	1010.1	1010.0
Medan	22	1010.5	1010.4	1009.6	1009.0	1008.6	1008.8	1008.8	1008.9	1009.6	1010.0	1009.8	1010.1	1009.5
Menado	16	1009.2	1009.4	1009.5	1009.0	1008.7	1009.0	1008.9	1009.1	1009.4	1009.1	1008.9	1008.6	1009.1
Padang	12	1010.0	1011.1	1009.7	1009.5	1009.2	1009.7	1009.9	1009.9	1010.4	1010.6	1010.3	1010.2	1010.9
Palembang	10	1010.1	1010.5	1009.8	1009.2	1008.9	1009.2	1012.9	1012.8	1012.6	1010.0	1011.4	1009.8	1012.9
Kupang	20	1008.4	1009.2	1009.1	1009.8	1008.5	1011.4	1012.0	1012.3	1011.7	1011.1	1009.5	1008.6	1010.3
Surabaja	20	1009.1	1007.8	1007.2	1007.7	1008.2	1008.6	1009.3	1009.3	1009.9	1009.4	1007.7	1008.2	1008.1

The general circulation over Indonesia is known as the monsoon; the wet and the dry monsoon. The names west and east monsoon are also used since the dominant winds are, respectively, westerly and easterly. They are currents which alternate from the west and the east, and occur within a period of one year.

In the wet season the eastern part of Indonesia is under the influence of a northeasterly wind coming from the north Pacific. This wind becomes northwesterly near the equator due to the coriolis force. In general, this bending does not occur exactly at the equator but somewhere to the north of it.

The climate of the western part of Indonesia is under the influence of an air mass from the winter anticyclone on the Asiatic mainland. This mass becomes northwesterly after having crossed the equator. This air is rich in water vapour due to the interaction of the Indonesian Ocean. According to the presently available radiosonde data from Djakarta, the relative humidity at 1,500 m is about 90%. This drops gradually to about 60% at 9,750 m in January. At 12,500 m, the humidity drops to practically zero. Only a few observations show values of 15%.

During the dry season, the southeasterlies from Australia dominate. The air is drier than in the other season, this being a consequence of the nature of the source and the path of the wind. An example of this is Ambon. This island is exposed to moist air due to the influence of the Banda Sea which causes abundant rain in this area. A zone of convergence may develop west of Irian Barat due to the topographical (mountainous) structure of this island (SCHMIDT, 1953). These facts might be the leading factors which cause rain in the southern part of Maluku (Moluccas) during the east monsoon.

The relative humidity for the above-mentioned altitudes are, respectively, 84% and 26%. These values are almost the same for 12,500 m during the wet season.

The values given above are to be reconsidered if more data becomes available in the future. By comparing data for both seasons a rough idea of the water content distribution may be obtained. It might be useful to mention here that in the dry season the 60% level is situated at 3,100 m. The difference in this value is very clear at the upper layers.

The upper wind over Indonesia has been treated by SCHMIDT (1952). Observations of

TABLE IV

MEAN RELATIVE HUMIDITY (%), 1951–1960

Station	Number of years	J	F	M	A	M	J	J	A	S	O	N	D	Mean
Ambon	14	74	74	76	65	83	83	84	83	81	78	76	76	79
Bandjarmasin	8	86	85	85	85	84	84	81	82	77	79	83	86	84
Balikpapan	8	85	86	87	88	86	88	85	85	84	84	85	85	85
Djakarta	28	85	85	83	82	82	81	78	76	75	76	76	82	76
Medan	22	85	82	83	83	82	82	84	82	84	84	85	85	83
Menado	16	87	86	86	85	86	84	79	76	78	80	85	86	83
Padang	12	80	80	80	82	80	78	76	87	89	82	83	82	80
Palembang	10	87	86	86	85	84	83	85	84	82	83	84	87	85
Kupang	20	85	85	85	81	68	78	75	74	74	75	79	84	79
Surabaja	20	84	83	83	84	82	79	77	73	72	71	76	80	78

00:00 h Z (07:00 a.m.) have been used to avoid disturbances caused by clouds and the influence of land and sea breeze circulation. The general circulation in Indonesia shows considerable variation between the same months of successive years. For this reason it was decided not to deal with the problem month by month, but by successive three month periods.

Since the purpose of this was to get a general picture of the circulation, local details were neglected. By using the graphical method as described by BELLAMY (1949), Schmidt obtained a result which gave the distribution of zones of divergence and convergence at various levels, especially the position of the Intertropical Zone of Convergence. The result might be summarized as follows, with special emphasis on the position of the Intertropical Zone of Convergence.

December–January–February

This period roughly covers the wet season in Djawa. The Intertropical Zone of Convergence occurs mainly over the Djawa Sea, and only occasionally over the island of Djawa during the course of January. But in the eastern part of the archipelago the air masses from the northern Pacific penetrate farther to the south, down to about 10°S. This difference in the location of the Zone of Convergence is obviously caused by the presence of vast mountainous regions in Kalimantan and Sumatra, which form serious obstacles to the monsoon current. This effect becomes less important with increasing altitude.

In general, the northerly stream changes its direction from northeast to northwest several hundred kilometers north of the equator, but not at the equator.

The intensity of the convergence increases with altitude up to 2,150 m. From there upward it decreases gradually. At 6,100 m the two converging currents cannot be easily identified because the whole of Indonesia is under the influence of broad, easterly winds. Over the southeastern part of Sumatra, strong convergence is found which is in accordance with the bad weather and very frequent thunderstorms usually occurring in this part of the country. Orographic influences are also important above and beyond the general effect of this Zone of Convergence.

The constancy distribution of the wind shows a high value over the Djawa Sea. This is caused by the fact that both converging currents are westerlies. Another high constancy is found over the southern Maluku. This means that the wind from the northern Pacific predominates over this area.

March–April–May

This is the transition period from the wet to the dry season. The position of the doldrums, on the average, is 3°S. In the lower layers of the atmosphere up to 3,050 m, the currents from both hemispheres curve toward the east near the Zone of Convergence. Consequently, the streamline pattern in this region is the same as in the wet season. This bending is accompanied by a strong decrease in velocity.

At 6,100 m the air streams are everywhere easterlies, except between Irian Barat and Australia where the circulation is anticyclonic. This might be due to the influence of the thermal high developed in the upper layers above the island of Irian.

In general, the intensity of the Zone of Convergence is less than during the preceding period. The same conclusion can be drawn for the constancies of the winds.

June–July–August

During the dry season period in Djawa, the currents from the Southern Hemisphere move over the equator to the Asiatic continent. At the lower levels, the bending to the north or northeast occur almost at the equator, whereas from 2,150 m upwards the shear occurs before the equator is reached. Roughly speaking, the Intertropical Zone of Convergance is situated north of Indonesia.

A strong cyclonic circulation is found to the south of Sumatra, particularly at 3,050 m. This might be due to the orographic influence of the mountain range of Sumatra.

The Zone of Convergence situated to the west of Irian Barat causes appreciable rains in south Maluku, as mentioned earlier. At the lower levels the vectorial mean wind velocities are about 7.5–10 m/sec in southeast Indonesia. These are somewhat stronger than during the other periods.

At 6,100 m the currents are easterlies over the whole area, except over south Sumatra where a slight curving to the north occurs. High constancies are found over the eastern part of the country. They are rather variable for the rest of the region.

At the lower levels, low values of constancies are found over the eastern and western part of the country, but high values are to be found over Kalimantan.

September–October–November

During this transition period the Intertropical Zone of Convergence is on its way back to the south. The streamline pattern is almost the same as in the preceding season, particularly in the southern part of the country. A less pronounced low pressure also develops to the south of Sumatra.

Finally, the current at higher levels up to 15 km is almost easterly throughout the year. The directions of the winds are between 57° and 114° with constancies of more than 50%. During the east monsoon, in particular, a value of 90% is reached.

Local winds

Due to the daily temperature range over large islands in Indonesia, the land and sea breeze is well established. This local circulation is enhanced by mountain and valley winds, especially along steep coasts. For this purpose, the island of Djawa is a good example. The mountainous structure of the country presents possibilities for the development of the föhn winds. The Bohorok is a local name for this kind of wind in Deli (near Medan in north Sumatra) which streams from the highlands of Karo down to the lowland.

The Bukit-Barisan Range plays an important role in this phenomenon. The wind streams up the west side of the range and the föhn effect takes place on the east side. In central Sumatra this wind is very famous in Padanglawas.

Tjirebon and Tegal (Central Djawa) are also exposed to such a wind during the dry season. In this case, the wind flows down from the western part of Mount Slamet and blows in a northern and northeastern direction. This wind is called the Kumbang.

In east Djawa, in the region of Pasuruhan and Probolinggo, a similair wind is named the Gending. In southwest Sulawesi there is also a föhn wind which is locally known as the Brubu. All these föhn winds occur during the east monsoon.

Climate classification

It has been recognized that in equatorial regions weather conditions undergo fewer variations than in temperate zones. However, this does not mean that there are no changes at all. Even the character of the seasons described above are not the same from time to time. In Indonesia, precipitation experiences the most important climatic variation. That is why any classification concerning the climate must stress the different behaviour of various areas with respect to the above-mentioned climatic element.

The latest publication dealing with this problem is by SCHMIDT and FERGUSON (1951). It shows that it is possible to apply the Köppen method, but that the results will not be entirely satisfactory. The authors suggest a new method called the Q system which takes the dry and wet months into consideration.

Köppen's system

This system is based upon temperature and precipitation.[1] It has already been mentioned that the monthly mean temperature in Indonesia does not change so much and, roughly speaking, remains almost constant throughout the year. However, distribution in the vertical direction plays an important role when dealing with this method of classification. The A climate (tropical, rainy climate) will predominate at lower altitudes. Then follow the C climates (warm, temperate, rainy climates) for higher altitudes. The border line between these two types is situated at about 1,250 m above sea level. Finally, the C climates change into the ETH (mountain) climate at about 3,000 m since, in general, monthly mean temperatures do not exceed 10°C above this elevation.

Since most places show mean temperatures of more than 22°C during the warmest month, the index "a" of the Köppen system will be used here. The Aa and the A climates are separated by an elevation of about 750 m.

In order to obtain a picture of the dry periods with their different characters, the A climates are divided into Af, As, Aw and Am. The As type is excluded, but the Am type is obviously important for Indonesia. The introduction of this last type is needed to indicate a tropical climate where the temporary dry period is not of great influence to the vegetation. A dry period is defined as a time interval at which, during at least one month, the mean precipitation taken over a number of years is less than 60 mm.

Since Indonesia belongs to the tropical region, the index "i" has been used to distinguish the C climate in this area from the temperate latitudes. This index indicates that the temperatures of the warmest and coldest months differ by less than 5°C. Also, the letter "h" appears as an index to show that the annual mean temperature exceeds 18°C. The elevation of 1,400 m is the border line between the Ci and Chi climates.

Furthermore, the C climates can be subdivided into Cw, Cf and Cs with respect to the character of the rainfall.

It is true that A climates will predominate over the lowlands, but there are small areas identifiable by a BS climate (steppe climate). The BW climates (desert climates) are not applicable for Indonesia. Examples of dry climates are places like Palu in Sulawesi and Waingapu in Nusatenggara (Lesser Sunda Islands). These are, respectively, BSi and BSwi climates. In general, a dry climate will be found only in Nusatenggara.

In this manner, Köppen's system may be used to construct a climatological chart for Indonesia. But, as SCHMIDT and FERGUSON (1951) pointed out, the results are not quite satisfactory. The reason for this conclusion can be found in the requirement of the Köppen system itself. This method is based upon mean values of precipitation for many years. In Indonesia, precipitation undergoes considerable variation from year to year. This is why an incorrect picture can be obtained. For instance; take the problem of dry months (less than 60 mm rainfall month). It might happen that the number of dry months derived from average precipitation equals zero, whereas the average number of dry months have a value higher than zero. Therefore, it is clear that Köppen's system does not have a practical purpose, for instance, for agricultural activities. On the other hand the second proposition gives a better picture, which makes it, of course, more useful in Indonesia. In 1950 a national working group was established to construct a climate map suitable

---

[1] For further details about Köppen's system, see pp.228 of this volume (Editor).

for practical application. This committee was formed on a broad base but, in general, the members consisted of meteorologists, agronomists, foresters and botanists. All interested parties had an opportunity to put forward their opinions and experiences. Finally, in 1951, the Q system, which will be discussed below, was developed.

Mohr's system

The Q system was intended to improve Mohr's method which, in turn, described the influence of precipitation and evaporation on tropical regions. Therefore, before describing the Q system, Mohr's system will be discussed very briefly.

MOHR (1933) divided the various months of the year into three categories; the wet, the moist and the dry months. This idea was based upon soil research.

Under his system, a month is considered wet if precipitation predominates over evaporation. This definition comes to a value of more than 100 mm/month for precipitation With a monthly precipitation of less than 60 mm, evaporation from the ground exceeds the amount of precipitation, so that the month is regarded as being dry. With a value of between 60 mm and 100 mm, there will be an equilibrium between evaporation and precipitation. In this case, the month is considered as being moist.

Furthermore, the climate type will depend on the number of wet and dry months. However, the procedure for obtaining this number is based on the monthly mean of precipitation averaged over many years. This averaging process is similar to the Köppen procedure, so that for the very same reason Mohr's system is not quite satisfactory. On the other hand, it has been demonstrated that this method has a broad application to agriculture on the island of Djawa. Therefore, it might be concluded that in Indonesia, Mohr's system is more successful than Köppen's system.

The Q system

This method uses the same precipitation limits as defined by Mohr in determining the character of the various months. The next step is to determine the number of dry and wet months year by year, and then to take the average value over many years. By doing so, the irregularities caused by the distribution of precipitation can be eliminated.

Finally, the climate type is determined by the quotient of the average number of dry months and the average number of wet months.

The climate of Indonesia is thereby divided into 8 types identified by the letters A to H with the following criteria:

$$
\begin{array}{lll}
A & 0 & \leq Q < 0.143 \\
B & 0.143 \leq Q < 0.333 \\
C & 0.333 \leq Q < 0.60 \\
D & 0.60 \leq Q < 1.00 \\
E & 1.00 \leq Q < 1.67 \\
F & 1.67 \leq Q < 3.00 \\
G & 3.00 \leq Q < 7.00 \\
H & 7.00 \leq Q
\end{array}
$$

The limits of these intervals will satisfy the equation:

$$Q = \frac{1.5\,a}{12 - 1.5\,a}$$

whereby, in the case of Indonesia, the symbol *a* is an integer from 0 to 8. Fig.1 is very useful in quickly determining the type of climate.

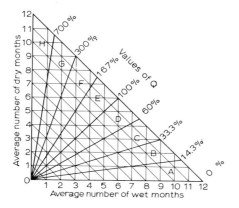

Fig.1. Diagram for the determination of climatic types according to the Q system. (After SCHMIDT and FERGUSON, 1951.)

Aside from this climatic division, the lines of elevation at 750 m, 1,250 m and 3,000 m are still important for agriculture purposes. The line 750 m is considered the border line between the lowlands and the mountainous area. As mentioned before, the region between 1,250 m and 3,000 m corresponds to the C climate according to the Köppen system and, finally, the 3,000 m line is the transition level from C to ETH climates.

Based on these wet and dry period ratios, and using the data for 1921–1940, the rainfall types of Indonesia might be summarized as follows:

Sumatra

Most of this island is the A type since the rainfall is of the equatorial type. The B type occurs mainly over the southern part on the east side and also over the mountain plains of Atjeh and Minangkabau. The north coast is identified by the C, D and E types, but this area is relatively small with respect to the whole island.

Djawa (Java)

The north coast of west Djawa belongs to the D type. South of this area is a band of type C and the rest is almost completely occupied by the A and B types. The A type region can be found in the central part, extended from south of Serang to east of Bogor and bending to the southwest of Bandung. Two small isolated areas of this type are also found to the north of Bandung and southwest of Tasikmalaja.

Central Djawa consists almost primarily of the B and C types. The B type is found over the central section and the south coast of the western part. Small areas of type A are situated to the southwest of Semarang and near Tjilatjap.

In east Djawa, the D type plays the most important role. This is followed by the C type.

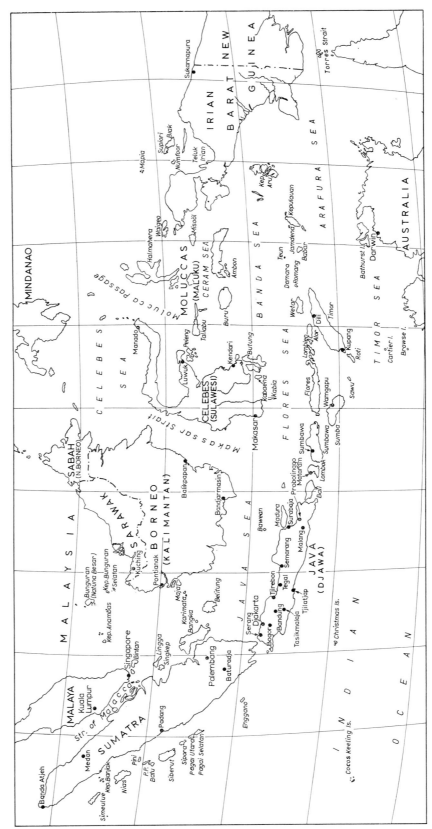

Fig. 2. Map of the Republic of Indonesia showing localities mentioned in the text.

226

A considerable broad band of type E appears along the north coast of the eastern part, with a small spot of type F at the centre.

Madura, Bali and Lombok

The island of Madura is almost completely covered by the D type. A section of the west coast has type C. On the other side, the east coast shows a considerable zone of type E. The centre of Bali, over the mountains, is marked by the B type. This is encircled by a ring of the C type and finally by one of the D type. The north coast is predominantly type E with a small area of type F.

A nucleus of type B appears at the centre of Lombok in the mountain range. This is followed by a circular band of type D. The north and south coastal zones are occupied by the E type. A remarkably broad band of type F is found along the east coast.

Nusatenggara (Lesser Sunda Islands)

This southeastern part of Indonesia is the driest region of the country. Isolated small areas identified by the B type are found in the mountains. All the coastal regions are occupied by the E and F types, with the exception of the northern part of the east coast of south Timor, and the west and south coasts of Sumbawa. The F types almost block the northern coasts of the large islands in this area. This F type is also dominant over the small islands situated to the east of Flores.

Consideration of this distribution from west Djawa to Timor reveals that the climate becomes drier and drier in the eastern direction. The reason for this phenomena has been mentioned in the previous discussion on rainfall; that is, the draining process of the water content as it streams from west to east.

Kalimantan (Borneo)

For the same reason as Sumatra, this whole region is almost completely under the influence of type A, with the exception of south coast and the southern part of the east coast of the island. These areas belongs to the B type. Further, no other types of rainfall have been found here.

Sulawesi (Celebes)

The dominant rainfall type over this island belongs to the A type. This type occurs over the central part, along the western part of the north coast and over a relatively small region stretching from the northern end southward over the mountains.

The south coast of north Sulawesi takes the C type. Only a small spot of D type appears at the centre of this coastline. The rest of the northern part is occupied by the B type. Central Sulawesi, as mentioned before, is mostly covered by the A type. B and C types are found over small areas in the middle and eastern part of this region. The area of Luwuk has the E type and is encircled by the D type. These last two kinds together with the F, G and H rainfall types may be found in the central highlands of Sulawesi, making them the driest regions in Indonesia.

TABLE V

RAINFALL TYPES BASED ON DRY AND WET PERIOD RATIOS
(After SCHMIDT and FERGUSON, 1951.)

Station	Height (m)	Number of years	Number of dry months mean	max.	<60 mm freq.	Number of wet months mean	max.	>100 mm freq.	Q (%)	Q system classifications	Köppen's classifications	Annual mean (mm) (After BERLAGE, 1949.)
Djawa												
Djakarta	7	20	3.3	6	2	6.7	9	2	49.3	C	Ama	1793
Bogor	266	20	0.3	2	2	11.5	12	13	2.6	A	Afa	4230
Tjipanas	1,070	20	0.9	3	2	9.3	12	2	9.6	A	Af	2817
Tjibodas	1,400	20	0.7	3	1	10.0	12	1	7.0	A	Cfhi	3380
Gunung Pangrango	3,023	20	1.3	3	3	9.5	12	2	12.6	A	Ethc	3324
Tjiandjur	459	20	1.4	3	3	9.3	12	1	15.0	B	Afa	2532
Lembang	1,300	14	2.3	4	2	7.9	11	1	29.1	B	Cfhi	2429
Indramaju	10	20	4.5	7	1	6.1	9	1	73.1	D	Awa	1621
Dawuhan	1,531	20	2.2	5	1	8.6	10	6	25.5	B	Cfi	3266
Djatibarang	25	20	3.2	6	2	7.7	9	5	41.5	C	Ama	2364
Gunung Butak	2,222	15	1.5	4	1	9.0	11	3	16.6	A	Cfi	3839
Baturaden	700	20	0.6	2	4	10.8	12	8	5.5	A	Afa	7069
Wonosobo	756	20	1.7	5	1	9.7	12	1	17.5	B	Af	4247
Jogjakarta	113	20	4.1	7	1	6.9	9	1	59.4	C	Ama	2181
Wonosari	210	20	4.3	7	1	6.8	9	1	63.2	D	Awa	1809
Semarang	2	20	2.5	5	1	8.2	10	3	30.4	D	Ama	2665
Surabaja	5	17	4.6	8	1	6.1	8	3	75.4	D	Awa	1775
Sarangan	1,290	20	3.4	6	2	7.7	10	2	44.1	C	Cfhi	2533
Probolinggo	10	10	6.3	8	4	4.9	7	2	128.6	E	Awa	1147
Djember	83	20	3.1	5	2	7.7	11	2	40.2	C	Ama	2296
Madura, Bali, Lombok												
Bangkalan	5	20	3.6	6	2	7.6	11	1	4.7	C	Ama	2024
Pamekasan	15	20	5.3	8	2	6.0	8	3	88.3	D	Awa	1583
Singaradja	40	20	6.5	9	1	4.4	6	3	147.7	E	Awa	1192
Tampaksiring	500	13	2.3	4	1	8.4	11	1	27.3	B	Afa	2666
Kintamani	1,475	16	5.5	8	1	5.1	7	2	107.8	E	Cwi	1964
Den Pasar	40	20	4.7	7	2	5.8	9	1	81.0	D	Ama	1737
Mataram	15	20	4.1	6	3	6.4	7	10	64.0	D	Ama	1802
Lombok	sea level	20	7.3	10	2	2.7	6	1	270.3	F	BSwi	726
Nusatenggara												
Bima	50	20	5.5	8	1	5.5	8	1	100.0	D	Awa	1286
Rutong	1,200	20	2.0	4	1	9.0	11	2	22.2	B	Af	3352
Waingapu	sea level	20	7.2	9	2	3.6	5	4	200.0	F	BSwi	768
Maumere	sea level	19	6.6	9	2	3.5	6	1	188.5	F	Awa	954
Kupang	2	20	7.2	9	1	4.2	8	1	171.4	E	Awa	1413
Kofamonanu	1,000	17	6.0	8	1	4.5	7	1	133.3	E	Aw	1226

Southeastern Sulawesi is mainly under the influence of the A, B and C types which respectively occupy the northern, the central and the southern part. The area of Kendari is marked by the D type.

The rainfall in southwestern Sulawesi is divided chiefly into two types. The western part belongs to the C type and the eastern side to the B-type. A small band along the south coast is covered by D, E and F types.

Maluku (Moluccas)

Consideration of the location of this group of islands which is spread out between Sulawesi and Irian Barat reveals that type B is dominant over this area. The central part of Halmahera and the southern part of Ceram are identified by the A type. The C type appears over the northern part of Buru and a small area over the northeastern part of Ceram.

Irian Barat (West New Guinea)

Due to its geographical location and mountainous structure, the A type is dominant over this region. In the southern part, it gradually becomes drier and drier which manifests itself in the formation of climatic bands ranging from B to E.

Table V presents the rainfall types for various places together with the corresponding classifications according to the Q and Köppen's systems. These places, with their various altitudes, are situated between the west point of Djawa and the island of Timor. The last column of Table V gives the annual mean of precipitation according to BERLAGE (1949). However, this column covers a period which is different from the one covered in the third column.

References

BELLAMY, J. C., 1949. Objective calculations of divergence, vertical velocity and vorticity. *Bull. Am. Meteorol. Soc.*, 30(1): 45.

BERLAGE, H. P., 1949. Regenval in Indonesia. *Verhandel. Meteorol. Geophys. Dienst, Batavia*, 37: 212 pp.

BRAAK, C., 1929. Het klimaat van Nederlands-Indië, I. *Verhandel. Kon. Magnetisch Meteorol. Obs., Batavia*, 8: 524 pp.

HAURWITZ, B. and AUSTIN, J. M. 1944. *Climatology*. McGraw-Hill, New York, 410 pp.

MOHR, E.C. J., 1933. De bodem der tropen in het algemeen en die van Nederlandsch Indië in het bijzonder. *Koninkl. Ver. Koloniaal Inst., Amsterdam, Mededeel.*, 31, deel 1.

PETTERSSEN, S., 1958. *Introduction to Meteorology*. McGraw-Hill, New York, 327 pp.

SCHMIDT, F. H., 1952. Upper winds over Indonesia. *Verhandel. Djawatan Meteorol. dan Geofis., Djakarta*, 45: 97 pp.

SCHMIDT, F. H. and FERGUSON, J. H. A., 1951. Rainfall types based on wet and dry period ratios for Indonesia. *Verhandel. Djawatan Meteorol. dan Geofis, Djakarta*, 42: 77 pp.

References Index

231

Geographical Index

Subject Index